The Accident Hazards of Nuclear Power Plants

The Accident Hazards of Nuclear Power Plants

Richard E. Webb

The University of Massachusetts Press
Amherst 1976

Library of Congress Catalog Card Number 75-37173
ISBN 0-87023-210-X
Printed in the United States of America
Second printing 1976
Figures 10 and 14 are reproduced through the
courtesy of Argonne National Laboratory.

Library of Congress Cataloging in Publication Data
Webb, Richard E 1939–
The accident hazards of nuclear power plants.
Includes bibliographical references.
1. Atomic power-plants—Accidents. 2. Nuclear
reactors—Accidents. I. Title.
TK1078.W42 621.48'35 75-37173
ISBN 0-87023-210-X

To my father and mother,
Clifford E. Webb and Betty Jane Webb

Contents

Acknowledgments

This author expresses gratitude to Professors David R. Inglis of the University of Massachusetts and Lynton K. Caldwell of Indiana University, and to Mr. Steve Gadler of Minnesota, and many others, who have enabled me to devote full time to my research. I am especially grateful to Professor Inglis, whose objective, scientific criticisms along the way were of great value, and to the many scientists, engineers and others in the nuclear industry, in the laboratories and universities, and in the Atomic Energy Commission and now the Nuclear Regulatory Commission who have been very helpful as well. I am also very grateful to William and Ann Carl of New York and the Natural Resources Defense Council of Washington, D.C., who have enabled me to participate in the proceedings of the AEC's Atomic Safety and Licensing Board, which proved very helpful in my studies, and to many other persons too numerous to name, though I wish I could.

Preface

THE SYNOPSIS which follows treats the hazards of nuclear reactor accidents. Treatment of this subject is not meant to suggest, however, that nuclear reactors pose the greatest risk of harm to the public and the environment, since it is difficult to assign relative risks. As A. DeVolpi of Argonne National Laboratory has well reminded us, there are other health and safety risks, and actual damage to the quality of life, that demand equal attention. He lists the risk of nuclear war; the risks of nuclear weapon accidents, including minor explosions that would disperse toxic plutonium radioactivity over the land; and the "most contaminating culprits," "chemical and biological waste discharges from municipalities, homes, and industry." * I would add motor-vehicle smog and noise, among other damages to the quality of life. In regard to conventional pollution, nuclear power can be a blessing; barring accidents and seepage of harmful amounts of radioactive waste materials into the environment, nuclear power plants could solve both the "energy crisis" and the air pollution problem in our cities. The energy resource in our reserves of uranium ore can last us about one thousand years, provided the "breeder" reactor is safe and practical; and since nuclear plants give off no noisome gases, such as the hydrocarbons and sulfur dioxide that are emitted by fossil-fueled power plants, they may well solve the air pollution problem. (However, nuclear power would have to replace fossil fuels, not just supplement them, if this benefit is to be

* A. DeVolpi, "Energy Policy Decision-Making: The Need for Balanced Input," *Bulletin of the Atomic Scientists* 30, No. 10 (Dec. 1974): 29–33.

realized.) Moreover, the electricity from nuclear power plants can be used to power mass transit systems and to make hydrogen gas, which might be usable as a substitute for gasoline fuel for motor vehicles. Since burning hydrogen gives off only steam (unless harmful nitric oxides are formed in combustion, as occurs with fossil fuels), the automobile smog problem could conceivably be solved by nuclear power. In addition, nuclear plants are compact and can be aesthetically pleasing relative to coal and oil burning plants.

Thus, it is clear that enormous benefits are potentially derivable from nuclear power. Nevertheless, we must carefully assess the hazards to establish whether the risks are acceptable. It may be that our safety and well-being will require nuclear power in the final analysis. But we have only just begun to perform the needed thorough evaluation of energy options and alternate ways of life, requiring consideration of economics, of pollution, of land resources, of national security, of whether the environmental impact of other technologies and industries needed to support nuclear energy will be acceptable, of preferred life-styles for pursuing happiness, of risks to safety and quality of life. Let us recall that the safety, well-being, and happiness of the People are the fundamental objects of society. To achieve them requires pursuing the truth, a difficult task, to say the least. This applies to nuclear power without exception; and it is in this spirit that I offer this analysis.

R. E. Webb
Amherst, Massachusetts
March 1, 1976

One
Introduction

NUCLEAR POWER plants present a hazard to the health and safety of the public because they are subject to *accident*, such as an explosion, in which harmful substances called radioactivity could be released to the atmosphere as dust and expose a large population to lethal or injurious radiation. The source of this accident hazard is the *nuclear reactor*, which is the heart of the plant. It generates the nuclear energy for making electricity, and in the process, it also generates the radioactivity as a by-product. This radioactivity builds up in the reactor and is even used as fuel in the case of plutonium, which is perhaps the most potent of all radioactive substances. In order to assess the hazards of nuclear power plants, therefore, we must investigate the accident potential of the various nuclear reactors in use or planned for use.

Several years ago this author, a nuclear reactor engineer, left the day-to-day work of nuclear reactor development to study full time and without constraints the accident hazards of nuclear reactors in all of the essential aspects. This study has concentrated on the present-day, water-cooled reactors in the United States (the pressurized water reactor [PWR] and the boiling water reactor [BWR]) and on the advanced "breeder" reactor (the liquid metal-cooled, fast breeder reactor [LMFBR]), which is the key element in the federal government's energy development plans. In 1973 he issued an interim treatise on the explosion hazard of the LMFBR, which raised a profound safety issue for that class of reactors.[1] He has now completed his study of the water-cooled reactors, and has also reexamined the LMFBR explosion hazard issue in the light of newer information. This book offers a synopsis of his research findings and analysis, including a highly critical review of the recent reactor safety studies of the Atomic Energy Commission (Rasmussen Report) and the American Physical Society,[2] both of which pertain to the water-cooled reactor, and of those sections of the AEC's *Proposed Final Environmental Statement* for the LMFBR program which deal with LMFBR safety.[3]

It is concluded, relative to the water-cooled reactors, that there is a good degree of conservatism in the industry's safety calculations to date and that extensive measures are taken to minimize the likelihood of serious reactor malfunction. However, there are many crucial questions and considerations concerning the validity of the safety calculations (the sufficiency of the conservatism), the overall potential for accidents and the resultant release of harmful radioactivity, the adequacy of the safety measures, and the probability of accidents, which have yet to be adequately addressed, according to this author's evaluation. Similarly, the LMFBR has an extremely serious uncertainty regarding a nuclear explosion hazard, not of the magnitude of an atomic weapon, though that has not been ruled out, but severe enough to release great amounts of harmful radioactivity.

The main conclusion presented herein is that the full accident hazard of each type of nuclear power reactor has not been *scientifically* established, even for the most likely of serious accidents. More specifically, the theory underlying the industry's safety calculations has *not* been *experimentally verified,* nor are the necessary experiments planned. The safety calculations in question are those theoretical predictions of the controllability of the most serious reactor accidents considered in federal government licensing, called the design basis accidents. Though a large number of small experiments are under way to test various aspects of theory, these are known to exclude essential physical processes that are crucial to verification. Adequate experimentation, then, requires *large-scale reactor tests,* primarily because of the peculiarities of the reactor physics involved. This shows up a fundamental shortcoming of the reactor hazards analyses developed by the nuclear community—the rejection of the *experimental philosophy* in favor of the use of *unverified theory,* or hypothesis, to build an "understanding" of reactor accidents. This shortcoming is one of the two chief concerns of this book.

The other, and more important, concern is that there are accident possibilities not considered for licensing which are more severe than the design basis accidents and that these have not even been theoretically investigated for the course they each could take, except for certain lesser cases; again, no substantive experiments are planned. In other words, the reactor containment systems for containing the radioactivity in the event of accidents are not designed for these worse accident possibilities, whose probability of occurrence is largely a matter of speculation. This lack of scientific knowledge of accident hazards is extremely critical, since each reactor appears to have an enormous potential for public disaster, as we shall see next.

Magnitude of the Hazard (The WASH-740 Report)

The theoretical magnitude of the worst consequences of the worst conceivable reactor accident is estimated below. This worst accident-worst consequences combination is not as unlikely as it might appear on the face of it. For such accident possibilities are very real, and their likelihood is ultimately a matter of personal judgment, as will be shown in the course of this book. The quantitative estimates of the worst conceivable reactor accidents are awesome and almost seem incredible; but they may be considered prudent in view of the limited scientific information that exists regarding nuclear reactor accident potentials. The author stands ready to change his evaluation, when and if the facts warrant it.

It should be added that the consequences can be much less severe under other, more probable, meteorological conditions than those assumed in the estimates. For example, rainfall is evidently needed in most cases to cause widespread radioactive fallout from a radioactive cloud created by a reactor accident, due to the washing-out effect of the rain. If there is no rain over the path of the cloud in the United States, the fallout or ground contamination could be minor. (In this case the fallout could occur over Canada or Europe, and though it would be more dispersed, more people would be affected.) For this synopsis, however, it is felt that the worst possible case should be presented, since it has not been adequately treated in previous studies by others. And, as one will be able to surmise from the estimates of the worst case, lesser consequences of a heavy release of radioactivity could still be severe.

To estimate the maximum consequences of any reactor accident, we adjust the estimates of the maximum possible reactor accident as given in the 1957 Atomic Energy Commission (AEC) report, *Theoretical Possibilities and Consequences of Major Accidents in Large Nuclear Power Plants* (WASH-740), to account for the sixfold increase in the highly intense, shortlived radioactivity and the fifteen-fold increase in the long-lived radioactivity in present-day reactors.* The maximum conceivable consequences of the worst accident are as follows: (1) a lethal cloud of radiation with a range of seventy-five miles and a width of one mile; (2) evacuation or severe living restrictions for a land area the size of Illinois, Indiana, and Ohio combined (120,000 square miles), lasting a year or possibly longer; and (3) severe long-term restrictions on agriculture due to strontium 90 fallout over a land area of the size of about one half of the land east of the Mississippi River (500,000 square miles), lasting one to several years, with dairying prohibited "for a very long time" over a 150,000 square mile area. There are other consequences not here estimated for water-cooled reactors, such as genetic damage. The potential accident consequences for the LMFBR—especially with respect to plutonium contamination, which may be a gravely serious lung-cancer hazard—will be discussed later, since they will depend on the explosion hazard unique to that reactor.

Incidentally, the maximum distance downwind from a reactor accident associated with the above land-area estimates of severe living restrictions and agricultural restrictions is about 1,500 miles to 2,000 miles. Hence, a nuclear reactor accident can affect distant communities as well as those nearby.

These extrapolations are not farfetched, since with much less radioactivity (for example, fifteen times less strontium 90) WASH-740 calculates up to 150,000 square miles of land requiring agricultural restrictions, 8,000 square miles of land requiring severe restrictions on living (an area the size of Massachusetts), and 680 square miles requiring total evacuation with a range of 118 miles. The WASH-740 figure for evacuation could easily engulf much of metropolitan Boston, for example, in the event of a reactor accident fifty to one hundred miles away, which would encompass presently planned locations. Moreover, the WASH-740 estimates were not meant to be upper limits, as worse weather conditions for promoting damages were not included: "[T]his study does not set an upper limit for the potential damages; there is no known way at present to do this." [4] Hence, the above estimates of the maximum accident consequences for present-day reactors, based on extrapolating the WASH-740 estimates, are not really upper limits either.

The above extrapolated levels of disaster will depend on

* Short-lived radioactivity is highly intense since it decays faster and therefore gives off a faster rate of harmful radiation. Long-lived radioactivity is a health hazard when it enters the body through the food chain and inhalation.

whether a large fraction (50%) of the radioactivity can be released from the reactor into the atmosphere upon an accident; and released in the form of a very fine, light dust, namely, dust particles one micron diameter in size, so that it can disperse over a wide area before falling out (one micron is about forty millionths of an inch). Whether such a release of radioactivity can occur will depend on the *reactor explosion potential,* which in turn will depend on the fuel temperature levels attainable in a reactor accident. This is because the radioactivity generated by the reactor builds up within the solid fuel material, so that in order for the radioactivity to escape the reactor, the fuel must overheat and melt or vaporize by a reactor accident. If the fuel melting temperature is high enough, most of the radioactivity would tend to boil, creating radioactive vapor (smoke). Fuel melting then enables the radioactive vapors to bubble out of the fuel and escape the reactor. Of course, if the fuel vaporizes, which occurs at higher temperatures, the radioactivity would vaporize right along with it. Furthermore, such hot, molten, or vaporized fuel could potentially cause an explosion, which could rupture the reactor enclosure and thus allow the escape of radioactivity (radioactive smoke) into the atmosphere. Hence, a higher fuel temperature will mean a greater fractional release of the radioactivity from the reactor, a finer radioactive dust, and a stronger explosion for expelling the dust.

Incidentally, in respect to fuel temperature, the fuel material in present reactors, uranium oxide, is more hazardous than the metallic uranium assumed as the fuel in the WASH-740 report, since the oxide fuel melts at a much higher temperature (5,000°F v. about 3,200°F), which means a greater chance for boiling off these radioactive substances.

In order to assess the risk of reactor accidents, therefore, it is necessary to examine the reactor explosion potential and the associated scientific uncertainties; and to question the practicality of an adequate reactor safety research program to resolve these uncertainties. Furthermore, it is necessary to assess the likelihood of serious accidents. For this task we need to appreciate the many different ways accidents can occur and the many subjective judgments inherent in any evaluation of reactor safety, and to analyze the experience of actual reactor mishaps and malfunctions.

Also, in order to assess the hazards, we will need to carefully analyze in detail the possible harmful consequences of a given heavy release of radioactivity to the atmosphere, including a social determination of what constitutes an "acceptable emergency dose" of radiation to the populace, an important part of the

WASH-740 report. We must try to foresee all possible effects and complications, such as the potential for atmospheric down-drafts that might swoop a radioactive cloud down onto a metropolitan area; the concentration of radioactivity in rainwater in roadside drainage ditches running along evacuation routes, which might conceivably cause lethal doses of radiation to a fleeing population; the potential for panic in evacuation; the public health burden on communities which must accommodate large numbers of refugees from contaminated areas; the potential for contamination of sources of drinking water, such as the Great Lakes; and the need for both shielding and the closing of windows, doors, and other openings to avoid high radiation exposures when taking shelter in advance of an approaching radiation cloud.

The present work, however, does not analyze in such detail the possible consequences of reactor accidents. It is believed that the first order of business is to analyze the potential for and likelihood of a heavy release of radioactivity in nuclear reactor accidents, since a heavy release would appear to be capable of causing intolerable disaster. This book, therefore, focuses primarily on the malfunction of nuclear power reactors, leaving the matter of the consequences to be thoroughly analyzed in another work, hopefully by other, more competent authority.

The next chapters will analyze the reactor accident potentials of the PWR, BWR, and LMFBR and will close with a few remarks concerning sabotage, nuclear energy centers, underground placement of reactors, and alternate reactors. Following the analysis of reactor accident hazards, the question—Who should decide whether reactors are safe?—will be considered. This question is believed to be essential because of the many subjective judgment factors that together will control any evaluation of the safety of nuclear power plants.

Before proceeding, the reader should become familiar with the terms Atomic Energy Commission (AEC), Nuclear Regulatory Commission (NRC), and Energy Research and Development Administration (ERDA). The AEC has recently been abolished and replaced by two agencies: the NRC, which is charged with the regulation of the health and safety aspects of nuclear power; and the ERDA, which is charged with promotion of nuclear energy, namely, reactor development. It will be convenient at times to continue to refer to the AEC, though the NRC or the ERDA will be implied.

Two

Present Nuclear Power Plants (Water-Cooled Reactors): Description and Accident Fundamentals

A WATER-COOLED reactor consists of a reactor vessel which contains a mass of fuel in the form of a large number of long, thin rods (see figures 1 and 2). The "fuel rods," numbering over 30,000, are arranged in bundles which are bunched vertically together to form the reactor "core." The individual fuel rod is a long, thin, metal tube in which small uranium oxide (UO_2) fuel cylinders are stacked (see fig. 3). The thin tubing (.025-inch tube-wall thickness) is the sole mechanical support for the 100 tons or more of UO_2 in the core and must remain intact if a disastrous reactor accident is to be prevented. (The tubing is usually called the fuel rod "cladding," since it clads the UO_2 cylinders.)

Water coolant is pumped through the core to remove the heat generated by the atomic reaction in the fuel rods for subsequent conversion to electric power. The coolant is very hot (550°F) and highly pressurized (1,100 psi and 2,100 psi for the BWR and PWR, respectively), and so it will explode into steam if the reactor vessel or connected piping should ever rupture. Reactor coolant pipes and valves are connected to the reactor vessel for the coolant recirculation process. In a PWR the coolant is 100% water (no steam bubbles), as the steam for powering the turbine-electric generator is made in the "steam generator" (boiler) outside the reactor. The hot reactor coolant flows through tubes inside the steam generator, and the heated tubes then heat the boiler water to generate steam. In a BWR the steam is formed in the core, so that at full power 43% of the core coolant volume consists of steam.

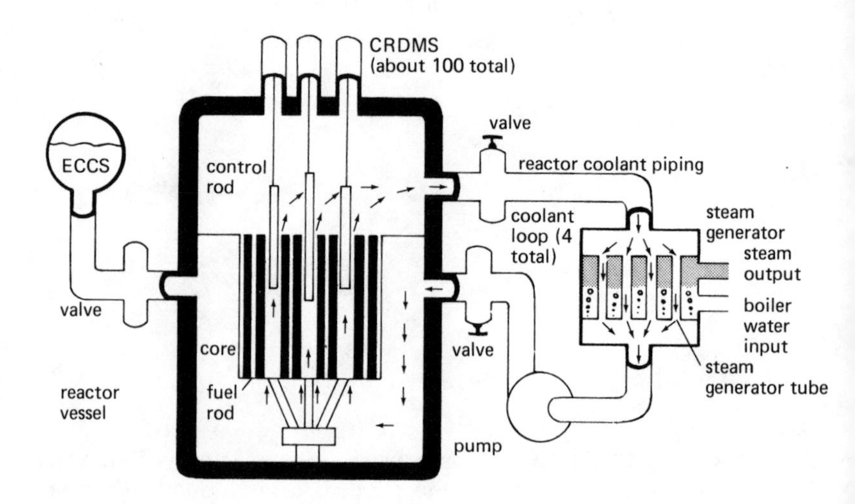

Figure 1. PWR system

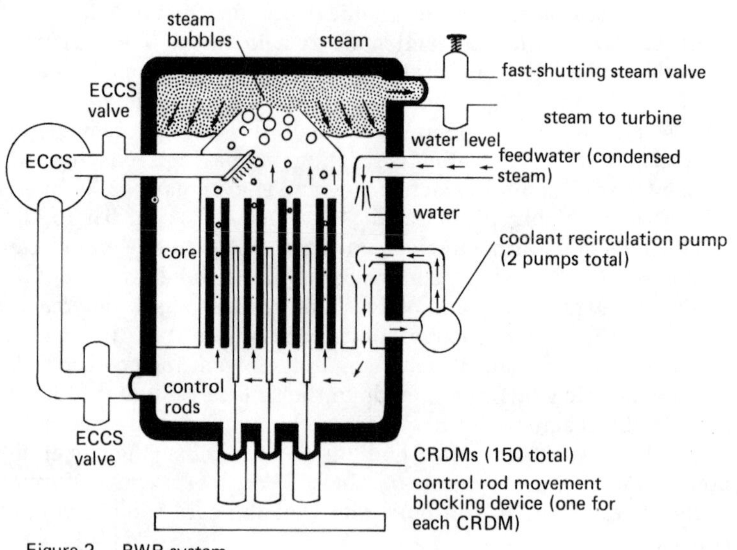

Figure 2. BWR system

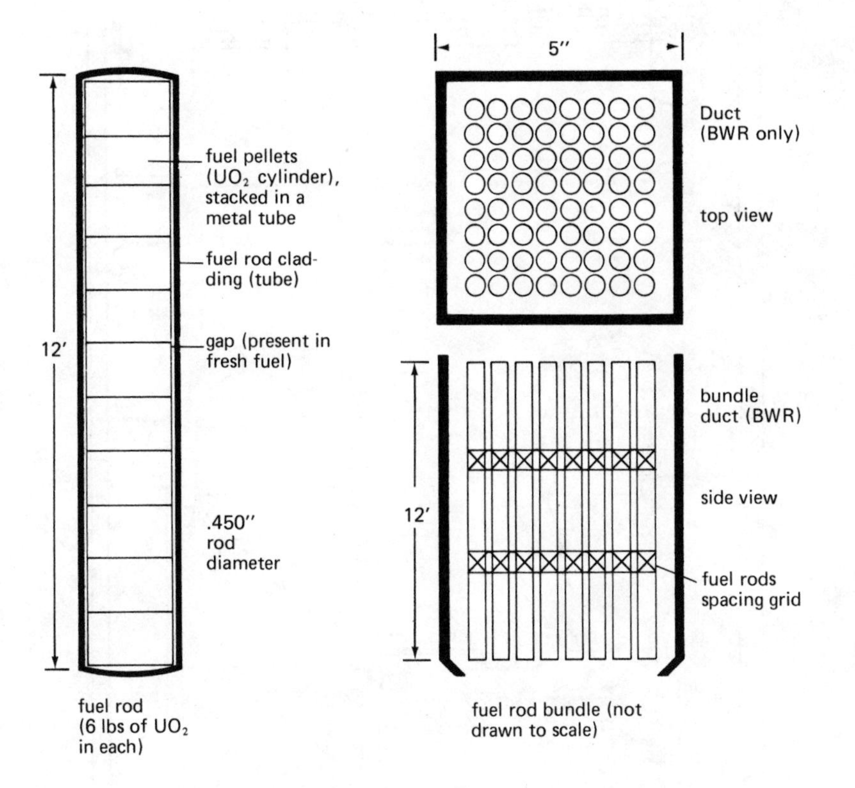

Figure 3. Fuel rods. Data: over 30,000 rods per core; over 500 fuel rod bundles per core; over 60 fuel rods per bundle.

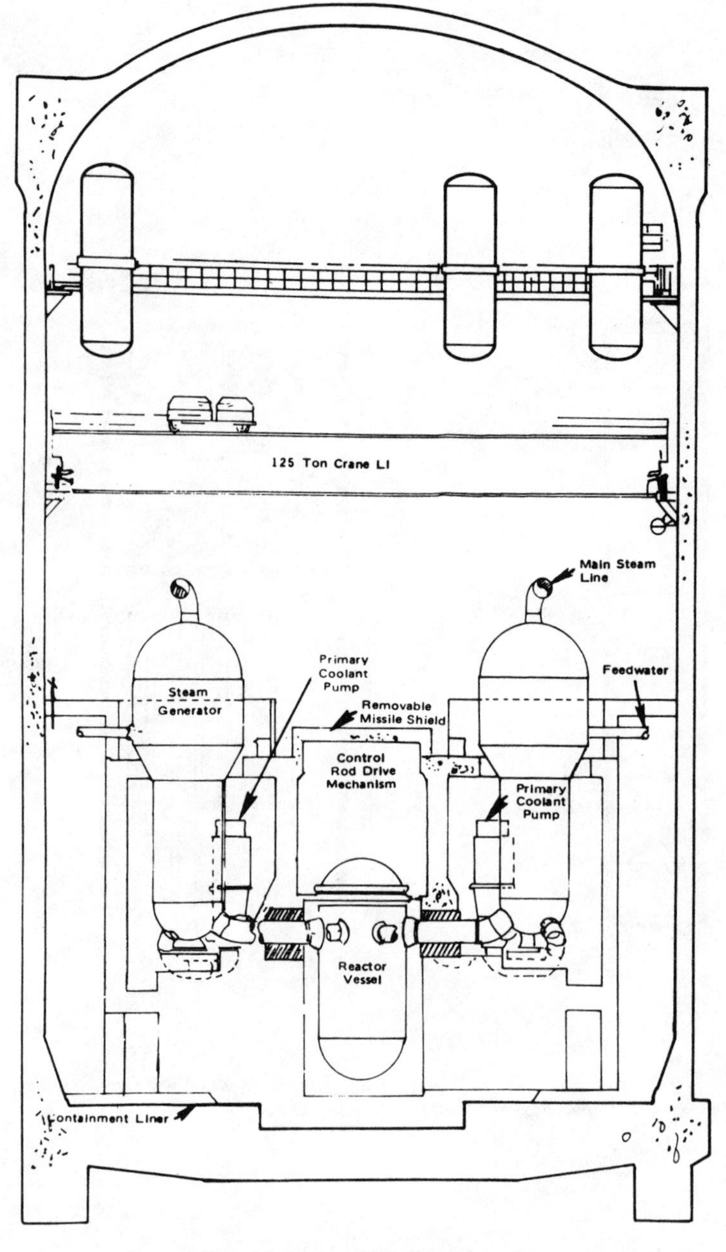

Figure 4. Typical PWR Containment

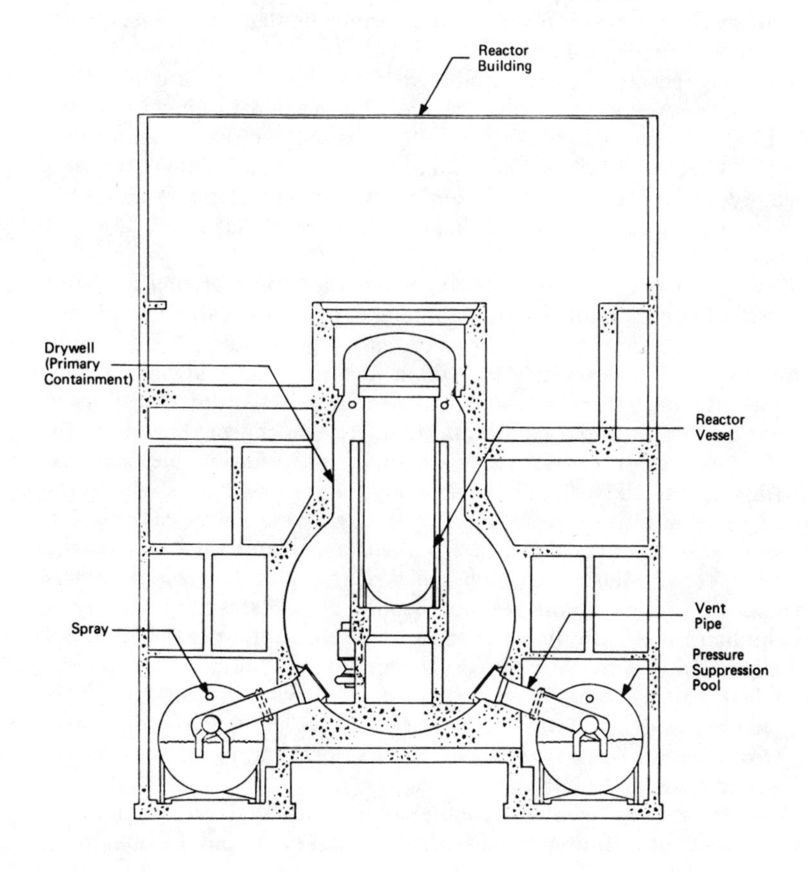

Figure 5. Typical BWR Containment

The reactor and its associated coolant system are housed in a large, leak-tight, reinforced concrete enclosure, called the reactor containment. Its purpose is to serve as a final protective shell or barrier against the release of radioactivity should the reactor coolant system rupture, *but only for certain limited accidents,* as we shall see (see figs. 4 and 5).

The reactor heat is produced in the fuel by an atomic chain reaction known as atomic fission. The "fission products," which thus build up in the fuel, are intensely radioactive, meaning that they emit harmful nuclear radiation, though their intensity gradually decays when the fissioning is stopped. This radiation is also a substantial source of heat in the core (about 1 to 7% of the fission power level).

The fission reaction occurs when the atoms of fuel are split by collisions with small subatomic particles called neutrons, which fly around inside the core, easily penetrating the fuel and cladding. For every fission, which absorbs one neutron, roughly three neutrons are given off, which in turn would cause three fissions in the next generation, then nine in the next, and so on, if the core were large and contained only fissionable material (this is the extreme hypothetical situation, which is given for illustration only; see fig. 6). The time between successive fission generations is very short in a real reactor, about thirty millionths of a second, due to the high speed of the neutrons, which means that the fission chain reaction would grow extremely rapidly if the neutrons multiply from one generation to the next. This rapid *exponential* growth process is known as "nuclear runaway," which can produce an explosion, or otherwise instant fuel overheating, and could thus result in a heavy radioactivity release. However, no nuclear-weapon-type explosion appears possible in a water-cooled reactor.

In a water-cooled reactor a controlled chain reaction is achieved by diluting the fissionable material * and by including rods of neutron absorber material, "control rods," in the core. In a PWR, the coolant normally contains dissolved boron, which is also a neutron absorber. These measures, plus the finite size of the core, allow a controlled, balanced rate of neutron losses, i.e., nonfission neutron absorption and leakage of neutrons from the core, such that only one neutron per fission will remain to

* Actually, natural uranium is already very diluted—.7% being fissionable (uranium 235 isotope) and the rest being practically nonfissionable (uranium 238)—and to achieve a controlled chain reaction in PWRs and BWRs requires the uranium to be slightly enriched in fissionable fuel, about 3%.

cause another fission. This balance maintains a constant fission rate, or reactor power level. In this balanced state the reactor is said to be "critical" (see fig. 7). Indeed, the fuel is so diluted of fissionable material that it cannot be made critical without the presence of water. This is because the neutrons, which are created at extremely high speeds, are slowed down by collisions with water molecules and because it happens that slowed-down neutrons are more effective in causing fissions.

However, if there should be a multiplicative imbalance in the neutron production-and-loss processes, measured by a quantity called *reactivity* (the percentage excess of neutrons per fission that will cause fissions in the next generation), the fission power level would grow or decay exponentially, depending on whether the reactivity is positive or negative, respectively.* An uncontrolled exponential growth would mean a nuclear runaway, as explained before; for though the neutrons are slowed down in a water-cooled reactor, their speeds still remain high after the slowing-down process (about six times the speed of sound), so that the time between successive fission generations is still short enough, thirty millionths of a second, to make a rapid nuclear runaway possible.

The reactivity must be varied by the reactor operator in order to raise or lower the reactor power level. It is normally regulated by slight, controlled movements of the control rods, upward or downward, by means of "control rod drive mechanisms" (CRDMs) located above or below the reactor vessel (see figs. 1 and 2). These CRDMS withdraw or insert the control rods out of or into the core, depending on the requirements of the situation. However, uncontrolled exponential power-level changes would occur even for slight reactivity changes were it not for the fact that a small fraction of the fission neutrons (about 0.7%) are released by the atomic fission fragments after one to ten seconds *delay;* these are called "delayed neutrons." The rest of the fission neutrons are released promptly at the instant of fission, and hence are called "prompt neutrons." Thus, if the reactivity—that is, the percentage excess of neutrons per fission—were less than the delayed neutron fraction, say 0.3%, the fission power level would rise, but at a slow, controllable rate, timed by the delayed neutrons, since there would be no excess of prompt neutrons per fission. Without an excess of prompt neutrons, the neutron multiplication process would have to wait on the fissioning that occurred several seconds earlier for a supply of excess

* The term *criticality* denotes zero or positive reactivity. The term *subcritical* denotes negative reactivity (decaying power level).

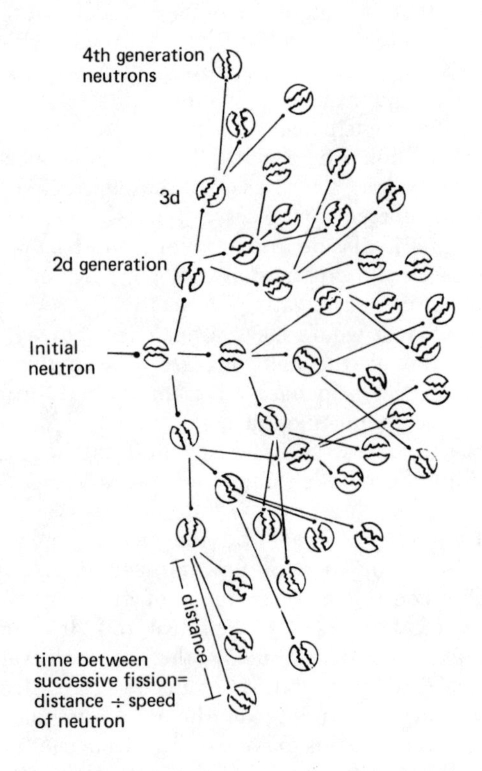

Figure 6. Nuclear runaway chain reaction (power excursion)

Exponential growth:

Number of generation, g	Number of neutrons N
1	3
2	3x3=9
3	3x3x3=27
4	3x3x3x3=81
.	.
.	.
.	.
g	$N=3^g$

Observe that g is an exponent of 3 in $N=3^g$. Such growth of the number of neutrons, N is thus called "exponential growth."

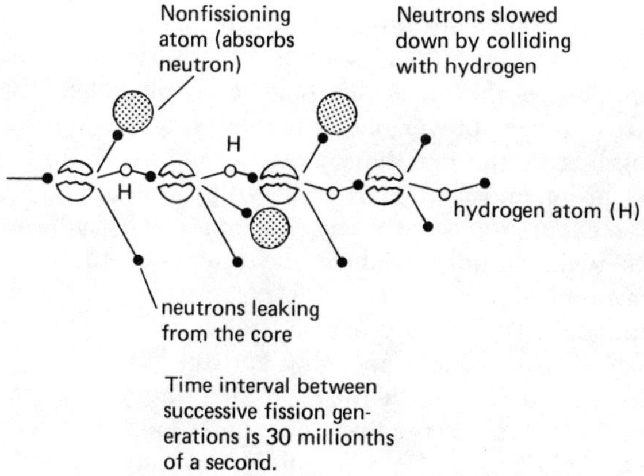

Nonfissioning atom (absorbs neutron)

Neutrons slowed down by colliding with hydrogen

H

H

hydrogen atom (H)

neutrons leaking from the core

Time interval between successive fission generations is 30 millionths of a second.

Figure 7. Constant fissioning chain reaction (reactor is "critical"). Note: The delayed neutrons are not depicted in this sketch. Also, there are extremely large numbers of fissions occurring per second in a reactor—about 3 billion trillion. The sum total adds up to the reactor heat output. Therefore, there is a like number of fission chains, as shown.

neutrons to increase the rate of fissioning. However, if the reactivity were to exceed the delayed neutron fraction due to some mishap, a nuclear runaway would occur, as there would be an excess of *prompt* neutrons per fission, which would mean that the neutron multiplication would proceed immediately from one fission generation to the next on the microsecond time scale, that is, in a rapid, runaway manner. Such a nuclear runaway is called a *power excursion* and results in a burst of power or fission heating in the core, which, again, can cause instant fuel melting or even explosively vaporize much of the fuel, say, 10%. (When the reactivity equals the delayed neutron fraction, the reactor is said to be "prompt critical," the threshold of nuclear runaway.*) The severity of an excursion increases as the reactivity increases. (We can now appreciate that the fact that no explosions have occurred in large nuclear power plants is due in part to our success so far in maintaining the reactivity below prompt critical, and not because nuclear power reactors are inherently stable.)

A power excursion is ultimately terminated by inherent reactor effects which quickly reduce the reactivity to below the delayed neutron fraction. These effects are called "negative reactivity feedback." The rapid reactivity reduction forces the core fissioning or power level to stop rising and then to decay rapidly to the preexcursion levels, thereby limiting the fission thermal "energy yield" of the excursion.† The main negative feedback mechanism is called the "Doppler effect," which normally will subtract reactivity promptly as the fuel heats up during a power excursion. However, the safe control of a power excursion by the Doppler effect will depend on the size of the initial reactivity increase being limited and on whether any *positive* reactivity feedback (autocatalytic) phenomena occur, which would worsen the excursion or cause secondary power excursions. Another negative reactivity feedback mechanism is coolant heating, as in boiling, which reduces the mass (density) of water in the core by thermal expansion due to the heat of the excursion. Since less water means less slowing down of neutrons, coolant heating increases the neutron speeds, which decreases the fission effectiveness of neutrons and hence decreases the reactivity.‡ For the same reason, a loss of coolant will shut down the fissioning. On the

* More fairly, the term *nuclear runaway* should be reserved for those strong power excursions that cause a reactor meltdown or explosion.

† Excursions last about 5/100 of a second.

‡ In some cases, coolant heating can increase reactivity in a PWR, namely, for high boron concentrations in the coolant, which could cause a dangerous autocatalytic effect (e.g., by reducing the boron density).

other hand, increasing the water density suddenly by injecting cold and unborated water into the PWR core, or by compressing steam bubbles in a BWR core, will increase the reactivity and, thereby, can cause a power excursion. It is clear, therefore, that the severity of a reactor accident will depend on the reactivity variations during the accident.* Hence, the reactivity is one of the most important quantities to calculate when predicting the course of postulated accidents.

In order to experimentally verify reactivity predictions, *full-scale* reactor tests are needed, since the reactivity is determined by the full assembly of all of the fuel and other materials and substances of a reactor core (and surrounding structure) together and its physical condition. All of the materials in and around the core affect the flow of neutrons within the core, leaving the core (leakage), and reentering the core (reflection back into the core), the net effect of which is to establish the neutron balance or imbalance (reactivity). Therefore, small-scale tests cannot reproduce the reactivity effects of large, full-scale reactors; and, hence, the results of small-scale tests cannot be extrapolated to full-scale reactors for verifying accident theory. This is an extremely important point, which will be drawn on repeatedly throughout this synopsis.

Generally, the reactivity can be affected (changed) by control rod movements, coolant pressure and temperature changes, fuel movements, fuel temperature changes, and boron concentration changes (PWR). For example, if the steam bubbles in a BWR core were suddenly compressed by a pressure surge, or if a control rod were suddenly ejected from the core, then the reactivity would rise dangerously. Or, if the core should melt down or explode apart completely, the reactivity would reduce to below zero and permanently shut down the fission reaction by core compaction, as in a meltdown, which squeezes out water, or by fuel "disassembly," as in an explosion. Moving fuel apart in a disassembly generally decreases the reactivity, by enhancing neutron leakage. (See table 1 for other examples.) To accurately predict the net reactivity change during the course of a reactor accident is extremely difficult, due in part to the complex interaction of many core processes and phenomena and due to the difficulty in rigorously calculating even the simplest of power excursions.

* However, the low fissionable material concentration, the slowed neutrons, the need for water to sustain the fissioning, and the Doppler effect all combine to ensure against a nuclear-weapon-type explosion in water-cooled reactors.

The basic protection against excessive reactivity and power excursions is the emergency reactivity reducing system, called the SCRAM, which rapidly inserts the control rods into the core automatically upon the detection of trouble by the reactor safety instruments.* The insertion of all of the control rods greatly reduces the reactivity to make the core strongly subcritical and, thus, to "shut down" the reactor.

The nature of steady fissioning is such that, unless the fuel rods are continuously "cooled" by the coolant, the fuel temperature will quickly rise without limit until the fuel melts within about twenty seconds. Thus, a loss of coolant is a way, other than a power excursion, to produce fuel melting.

Whether the effects of initial fuel melting will at least shut down the fission reaction—or cause secondary power excursions through autocatalytic reactivity effects—is a major uncertainty. Even if the fissioning is stopped, however, there is still the "afterheat" from the decaying fission product radioactivity, which is substantial—about 1% to 7% of the reactor full-power level, depending on the circumstances of the accident. This afterheat, too, must be removed by water coolant, or else the fuel will heat up and eventually melt in three to twenty minutes.

Table 1 Core Changes Affecting the Reactivity (Water-Cooled Reactor)

Increase reactivity	Decrease reactivity
Control rod outward movement	Fuel heating (Doppler)
Steam bubble compression or collapse (BWR)	Coolant heating
Boron reduction (PWR)	Coolant boiling
Inward fuel rod bowing	Coolant pressure decrease
Coolant temperature decrease	Increased boron concentration
Fuel temperature decrease	Control rod insertion
Coolant pressures rise	Fuel compaction as in meltdown (which squeezes water out of the core)
Coolant addition	Coolant loss
	Core disassembly, such as explosion

* The term SCRAM has its origin in the original reactor experiment in which the operators were instructed to reduce the reactivity and then quickly leave the premises ("scram") in the event of a power excursion.

The uranium oxide fuel melting temperature is 5,000°F, which is 1,500° hotter than the core of a steel blast furnace. This extreme temperature could boil a large fraction of the fission products out of the core for eventual injection into the atmosphere, and also could result in very strong *steam explosions* (explosive boiling of water), should the molten fuel subsequently come into contact with pockets of water in the system. Moreover, the metal tubing, being zirconium, would at those temperatures chemically burn in water to generate still more heat,[1] and to form hydrogen gas, which could itself ignite outside the reactor vessel in an explosion. Steam explosions, hydrogen explosions or hydrogen gas burning, or the gas pressure of the hydrogen alone could rupture the containment, according to theoretical-limit-type calculations.[2]

To ensure that the fission heat or afterheat is safely removed, the fuel rods must remain intact for adequate coolant contact; that is, they must not crumble, for it is theoretically conceivable that a crumbled, uncoolable mass of fuel, even if only just one thousandth the size of the core (200 pounds), could heat up to the melting temperature in four to twenty-five minutes due to the afterheat alone, depending on the afterheat level—or in about twenty seconds, if the crumbling occurred at full power. Furthermore, it can be calculated that 1% to 3% of the core mass, if molten, could cause a steam explosion to rupture the reactor vessel,[3] which presumably would destroy the entire core within it. Hence, the steam explosion force caused by a 200-pound molten fuel mass could conceivably crumble 5% of the core around it, which in turn could heat up and melt, and then cause a catastrophic steam explosion that ruptures the reactor vessel. This could then result in a loss of all coolant, followed by core-wide meltdown. Such a cascading core meltdown and explosion process can only be assuredly prevented by avoiding fuel-rod crumbling in the first place.

Three
The Classes of Reactor Accidents

THERE ARE four classes of reactor accidents in which the fuel could dangerously overheat: loss-of-coolant accident (LOCA), spontaneous reactor-vessel-rupture accident, power-cooling mismatch accident (PCMA), and the above-mentioned power excursion accident (PEA).

In a LOCA, a coolant pipe is assumed to rupture spontaneously. The hot, pressurized coolant would then blow itself out of the reactor, flashing to steam as the coolant pressure is relieved. This process is called the coolant "blowdown." The basic protection against excessive overheating of the fuel in a LOCA is the Emergency Core Cooling System (ECCS), which injects auxiliary coolant into the reactor following the pipe rupture (certain types of LOCAs also require a SCRAM). Although fuel melting and crumbling are predicted to be avoided in a LOCA that is controlled by the ECCS, a substantial amount of radioactivity could escape from both the core (fuel rods) and the reactor vessel, if fuel-rod cladding ruptures occur, which are conservatively predicted.

A much more severe loss-of-coolant accident would be a spontaneous rupture of the reactor vessel, due to a design or manufacturing defect. The ECCS would be ineffective in this situation, since the core could not be reflooded, except for limited vessel ruptures above the core level. Moreover, in the extreme case, the reactor vessel closure head, if blown off, could become a missile and easily pierce the containment.[1] Though in one sense the worst vessel rupture is a LOCA, its immediate containment-rupture effect makes it follow a qualitatively different course. That is, the core meltdown would *follow* the containment rupture; and, thus, the boil-off of fission product radioactivity would have a direct access to the atmosphere. In contrast, the containment-rupture/core-meltdown sequence would be reversed in a LOCA-without-ECCS, which could allow much or most of the fission products released from the reactor to be absorbed within the containment by condensation, and so on, before the containment is ruptured by the effects of core meltdown. Thus, a LOCA-without-ECCS would have a reduced potential for radioactivity release than a vessel rupture, which is why the two accidents are distinguished.

A PCMA (power-cooling mismatch accident) occurs whenever a region of the core overheats due to excessive fissioning or undercooling, when the reactor is operating at around full power. A PCMA conceivably could result in a cascading core meltdown, followed by steam explosion and rupture of the reactor vessel or coolant piping—or conceivably could even generate a severe power excursion, by causing sudden changes within the core that raise the reactivity.

Severe power excursions (PEAs), sometimes called "reactivity accidents," could cause the core to melt more rapidly than a PCMA, followed by a steam explosion and vessel rupture—or could cause instant explosion and, of course, vessel rupture. Note that catastrophic PCMAs and PEAs are more severe than a spontaneous reactor vessel rupture, since the fuel and coolant would be hotter and at higher pressures at the time of the reactor vessel rupture.

There are a great many different possibilities for PCMAs and PEAs. Some of them are protected by a fast insertion of the control rods to shut down the fissioning, i.e., a SCRAM, which must occur automatically within one second of detecting trouble. Others are protected by the opening of "relief valves" to relieve the pressure of overhead coolant, while the reactor is shut down by other means. However, the worst PCMAs and PEAs cannot be controlled; and, therefore, their probability of occurrence is (hopefully) made remote by careful operation, maintenance, and preventive devices, such as interlocks, and so on.

Finally, there is the *reactor containment* for containing the steam in a LOCA. By containing the steam, any radioactivity that may have escaped the reactor and mixed with the steam will also be contained, except for minor seepage. Water sprays inside the containment (PWR) help quench the steam and wash out airborne radioactivity from the containment air. BWRs use pools of water for quenching the steam. However, the containment is designed only for a pipe rupture (LOCA) in which the ECCS is successful in preventing fuel melting. The containment is *not* designed to contain, or withstand without rupturing, a LOCA with an ECCS failure, a reactor vessel rupture, *or uncontrolled PCMAs and PEAs* in which a core meltdown or a prompt, explosive power excursion occurs.[2] However, if the containment suffers only a limited crack and if the containment water sprays operate (PWR), most of the airborne radioactivity released from the core and into the containment atmosphere could be washed down or otherwise confined to within the containment, and the steam pressure eventually quenched, to avert a heavy radioactivity release to the earth's atmosphere.[3] This will depend on limiting the severity of the reactor accident, for it is theoretically possible that either a steam explosion or a hydrogen explosion can completely rupture the containment, even if only 10% of the core became molten and interacted with water.[4]

For completeness, we should add the possibility of the containment simply being open (a door or hatch left open inadvertently) when a core meltdown or explosion occurs, which would allow the radioactivity released from the core to escape to the atmosphere without the need for a forced breach caused by the reactor accident. This should always be kept in mind.

Having described the reactors, their accident modes and safety features, we can now proceed to examine some of the essential reactor accident hazards and uncertainties.

Power Excursion Accidents

To begin with, though very serious, the LOCA, short of a spontaneous reactor vessel rupture, is *not* the most serious reactor accident, as the public has been led to believe. Rather, the PEA is by far the most worrisome one, with the PCMA somewhere in between. This is because, if the ECCS fails to operate or is ineffective in a LOCA, the core heat-up will still be a relatively slow process compared to a PCMA or a PEA, since only the less intense afterheat will be present. (In a LOCA, the fissioning is predicted to stop shortly after the pipe rupture, due to the negative reactivity feedback of the loss of coolant.) A slow heat-up minimizes the chance among the three accident modes for severe steam explosion. It also minimizes the degree by which the core temperatures could exceed the 5,000°F UO_2 melting temperature, and this in turn minimizes the fraction of radioactive fission product release from the reactor. Furthermore, a slow heat-up maximizes the fallout of the radioactive fission products within the containment and washout by the containment sprays (PWR), since the containment and sprays (PWR) will function for a relatively long time (about an hour or so) before a core meltdown and the possible resultant steam explosion, hydrogen explosion, or overpressure could occur to rupture the containment. Nevertheless, a large scale (50%) fission product release following a LOCA has not been scientifically ruled out, as will be discussed later.

In contrast, severe PEAs have a theoretical potential for rapidly achieving higher fuel temperatures and a stronger steam explosion, which could instantly rupture (explode) the reactor vessel,[5] compounded by the sudden, explosive release of the hot, pressurized coolant and destructive missiles (flying fragments of the reactor vessel).[6] Thus, severe PEAs would more likely rupture the containment completely, and more rapidly, than the LOCA-without-effective-ECCS, allowing little, if any, time for in-containment washdown and fallout of fission products, assuming the sprays (PWR) were not knocked out. Therefore, of all reactor accidents, the worst possible PEAs may cause the maximum release of fission product radioactivity.

Moreover, the ECCS is likely to prevent core melting in the event of a LOCA, if the circumstances of the LOCA are confined to those which the ECCS is designed to control, namely, a pipe rupture occurring when the reactor is being operated steadily at its full power level.[7] This narrowly defined LOCA is called the "design basis, loss-of-coolant accident" (DB-LOCA), and is the focus of attention by popular critics of nuclear power.[8] This likely success of the ECCS is due to the fact that the *afterheat,* which the ECCS may need only control in a DB-LOCA, is only a small fraction of the full reactor power (heat generation) level and is well predictable with a definite decay rate. This allows the ECCS to be designed with considerable confidence and definite, substantial "conservatism," that is, safety margin. However, the predicted ECCS performance still has not been verified experimentally; and no adequate program is presently planned, as this will evidently require LOCA tests using large- or full-scale reactors, which are not planned. (This will be discussed later.) Though formidable, an adequate LOCA-ECCS experimental program might be practical, as we shall see. In contrast, the state-of-the-art calculations of the heat source in a PEA, namely, *the fissioning,* are not well founded and may not even be practical for severe PEAs. Furthermore, to experimentally verify the PEA theoretical predictions would require destructive power excursion experiments using full-scale reactors, because of the *reactivity*,[9] which may be impractical. (This, too, will be discussed later.)

The upper limit of PEAs is set by a quantity called "excess reactivity," which is the maximum reactivity level the core could theoretically reach if all of the control rods were instantly withdrawn (ejected) from the core when the core is filled with water. The excess reactivity is about 16% when the reactor coolant is at operating temperature; [10] whereas a reactivity of only 2% should produce a reactor explosion.[11] In a BWR, the reactivity increase potential of a single ejected control rod can be as high as 5%.[12] (Recall that the threshold for power excursions is when the reactivity equals the delayed neutron fraction of only about 0.7%.) Therefore, the reactivity potential exists for disastrous explosion, which is unavoidable, as the excess reactivity is needed to offset the gradual reactivity loss that occurs with fuel burn-up. (The control rods are gradually withdrawn from the core, and the boron is diluted [PWR], as the fuel is consumed by fissioning over a period of a year or so. Thus, as the fuel burns up, the excess reactivity is slowly consumed.)* If a PEA does not produce instant melting and explosion, it conceivably could at least cause a substantial amount of fuel to crumble into an uncoolable mass, which could then cause progressive core melting and steam explosions. Such explosions could cause the reactor vessel to rupture or even cause secondary power excursions by, for example, rupturing several CRDM (control rod drive mechanism) housings and ejecting the associated control rods.[13]

Examples of severe PEAs are as follows: [14]

PWR or BWR control rod ejection accident. The pressure housing of a control rod drive mechanism suddenly ruptures and blows away from the reactor vessel; the associated control rod is then blown out of the core by the high pressure coolant, which increases the reactivity. In a BWR a device is normally provided to block the control rod movement in the event of a CRDM housing rupture, to prevent rod ejection; but this device could be ineffective, due to improper installation—or it may not have been installed (see fig. 2). Normally, the reactivity potential or "worth"

* In certain circumstances, the excess reactivity is still substantial at the end of the "life" of the core, so that severe PEAs are still possible then. Otherwise, the excess reactivity would be zero at the end of life. One such circumstance is when the coolant is cold; cooling down the coolant raises reactivity.

of the control rods in a PWR is much less than that of the BWR, due to the boron in the PWR coolant, which offsets most of the excess reactivity, allowing the PWR control rods to be nearly withdrawn during reactor operations and thereby minimizing their reactivity worth for causing a power excursion.

Cascading control rod ejections. Should a faulty CRDM housing break off, the associated violence could conceivably trigger several adjacent CRDMs to break off in a rapid cascade manner, assuming they are similarly faulty, causing successive reactivity increases.

BWR control rod drop out accident. A control rod becomes detached from its drive shaft and falls out of the core, raising the reactivity.

The control rods are normally positioned in the core in such a way as to minimize the reactivity potential of a single control rod accident. Thus, for some of the PEAs to be severe, a violation of the normal positioning would have to exist concurrent with the control rod malfunction (ejection or drop out), in order to produce a strong reactivity increase. Such violations are nevertheless possible and have already occurred in near-accident incidents (see p. 180 and app. 2, no. 7).

BWR steam valve closure accident. While a BWR is being operated at full power, the steam valve suddenly trips shut, stopping the reactor steam from flowing out of the reactor (see fig. 2). The steam valve is purposely set to trip shut in order to stop powering the turbine-electric generator systems in the event of their malfunction, which is not uncommon. Steam would continue to be generated in the core, however, since the core fission power level would remain high. The result is that the steam would rapidly accumulate in the reactor vessel, causing the steam-coolant pressure to rise. This rising coolant pressure would compress the steam bubbles in the core, which in turn would raise the reactivity to trigger a power excursion. This situation would normally be protected by the SCRAM (rapid insertion of the control rods), which would shut off the reactor fission power by rapidly reducing the reactivity; but it would become an accident if the SCRAM failed to occur. In the latter case, the power excursion would be momentarily halted by the negative Doppler reactivity effect of the prompt fuel heating. But due to the SCRAM failure, the core power level would continue to rise beyond the full power level, since the pressure rise would continue to increase the reactivity.

The reactor is equipped with steam relief valves, commonly known as safety valves, which are designed to pop open when the coolant is over pressurized; but the relief capacity of these valves is not large enough to prevent the pressure from rising if the core power level is not reduced.

For this accident, a back-up measure for reducing the reactivity is provided in the event the SCRAM system fails to function. This consists of an automatic turn-off of the coolant recirculation pumps, which slows down the coolant flow through the core. This causes the coolant to boil more, thereby offsetting the steam-bubble-compression-reactivity-rise process. Therefore, this accident would become much worse if both the SCRAM system *and* the turn-off system for the recirculation pumps should malfunction—which could happen independently and lie hidden until the steam valve trips shut.

Figure 8 shows a calculation by Brookhaven National Laboratory of the core power excursion for the case of SCRAM failure, but assuming the coolant pumps are automatically turned off.[15] The worse case of the added failure of the coolant pump stoppage protection has been calculated by this author and is also shown in the figure for comparison. The worse case leads to fuel melting, and a more dangerous situation, which will be discussed shortly.

PWR cold water accident. This assumes that one of the three or four coolant loops has been closed off from the reactor by the valves of the loop and that the coolant in the closed loop piping is cold and free of boron. A sudden injection of the cold, unborated water into the core from the off-loop by an abrupt opening or failure of the two loop valves, when the reactor is operating with the remaining coolant loops, would then cause the reactivity of the core to increase greatly. (However, about 60% of the PWRs do not have coolant valves, which eliminates this source of PEA for those PWRs.)

PWR core drop accident. This extreme PEA is caused by the heavy, 100-ton core barrel breaking away from its support and falling, despite the back-up supports, which are assumed to fail as well. Since the control rods would remain stationary, because they are held by the CRDMS on the vessel head, the core by falling would in effect cause all of the control rods to move out of the core together. This would presumably cause a very severe PEA—though it would depend on the distance the core could fall, which could not be determined from available information. A core drop might be caused by an earthquake, for example, or by a mechanical failure.

Autocatalytic reactivity effects. Many disastrous PEA possibilities involve a strong reactivity increase caused entirely by a single source of reactivity rise. However, compounding the risk

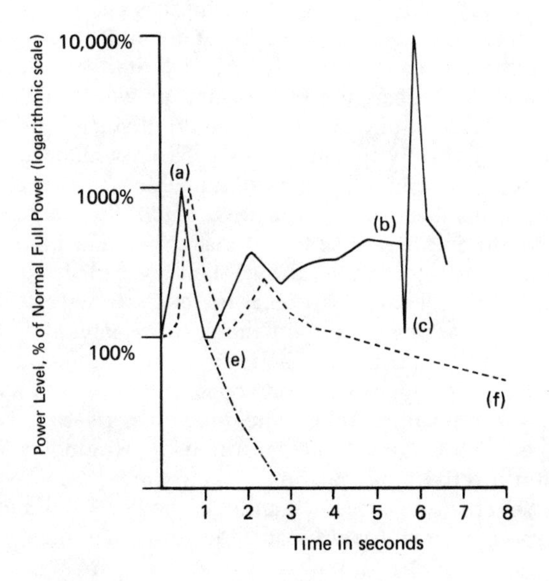

Figure 8. BWR steam valve closure accident without SCRAM and assuming that coolant recirculation pumps are not turned off. (Core power level versus time)

a. Power excursion caused by steam valve closure
b. Fuel melting
c. Power level drops momentarily due to negative reactivitity effect of the initial fuel melting and discharge.
d. Catastrophic Nuclear Runaway (Autocatalytic)—Results in core explosion, 8,000° F peak fuel temperature. Autocatalytic reactivity effect due to the "Roman candle" effect upon initial fuel melting. (Assuming reactor pressure rises 500 psi in one second, due to Roman candle effect.) Source: Author
e. Power level if the SCRAM occurs.
f. Dashed curve shows results of Brookhaven National Lab. calculation. They assume that the coolant pumps are turned off (no SCRAM). No fuel melting then occurs, and reactor power is then controlled. Source: BNL 17608, fig. VIII-5.

of PEAs is the possibility for *autocatalytic reactivity effects* occuring in response to the effects of an initial power excursion, or other disruptive events, which could draw on the excess reactivity and feed the core a strong dose of additional reactivity.

Conceivable examples of *autocatalysis* are (1) inward bowing of fuel rods,* should certain mechanical supports (fuel rod spacing grids) be faulty [16] or ineffective (if the fuel rods break into pieces and shift around in a PEA); (2) compression of steam bubbles in the core by the pressure surge of a mild steam explosion; (3) prevention of control rod SCRAM by fuel distortion; (4) secondary control rod ejections; (5) sloshing of water coolant in the core of a BWR at power, which could displace steam bubbles and raise the reactivity rapidly—a process which could occur in an earthquake or in any accident in which the reactor vessel is bounced or shaken; and (6) positive reactivity feedback due to coolant heating in a PWR. This last example might be possible if the boron concentration in the coolant exceeded safety limits due to some fault. Then, if an increase in the power level occurred (however slight), the resultant increase in coolant temperature would cause an increase in reactivity, which would make the reactor power rise faster and heat the coolant even more, and so on in an unstable fashion. When the reactivity rises above prompt critical, an autocatalytic nuclear runaway would occur. Such autocatalytic reactivity effects would make the initial power excursion more severe, or generate secondary power excursion, to further increase core temperatures and the force of the explosion.

Figure 8 shows the results of one calculation performed by this author of one such conceivable autocatalytic effect occurring in the above-mentioned BWR *steam valve closure accident* (without a SCRAM and without stopping the coolant pumps). Conceptually, this process occurs as follows: The power level rises after the initial power excursion, since the steam pressure continues to rise. In about six seconds fuel melting occurs. The fuel melting will first occur in the hottest region of the core, which will involve only a few fuel rod bundles. The outpouring of molten fuel from the rods, and its immediate contact with the water coolant, produces vigorous boiling, a coolant pressure surge, and expulsion of the molten fuel-water-steam mixture. The long, open-ended metal box or duct surrounding each fuel bundle, called a "coolant

* A very substantial autocatalytic reactivity effect due to fuel rod bowing was actually observed in a mild power excursion experiment of a small core; see n. 16.

channel," constrains the fuel-water-steam mixture so as to discharge it out of each end, much like a Roman candle," * into the coolant region above and below the core to cause still more boiling and a strong pressure surge (see fig. 9). This pressure surge, assuming it happens very quickly, would compress the steam bubbles in the remainder of the core, where the fuel has not yet melted and poured into the coolant. The bubble compression would increase the reactivity, following a momentary loss of reactivity due to the molten fuel discharge,[17] and could thereby produce a very strong, final power excursion that would be catastrophic. The final power excursion is calculated to make the fuel in the core molten on the average, with much of the fuel itself vaporized to explosive pressures, which theoretically could produce a steam explosion that would easily rupture the reactor and the containment.† The coolant pressure at the time of such explosion would be twice the normal operating pressure, which would compound the explosion.

It may be that consideration is now being given to increasing the steam flow capacity of the steam pressure relief valves to prevent the pressure from rising in the first place and thereby avoid the autocatalytic power transient. But whether this will be done remains to be seen and justified by analysis. Increasing the relief valve capacity may have adverse effects on other things, which would have to be considered. For example, having more relief valves raises the chances for stuck-open valves in overpressure situations, which could allow enough coolant to escape the reactor to threaten uncovering the core, leading to a possible core meltdown, unless the ECCS functioned to replace the lost water. Then, too, there is always the possibility of relief valves not working.

Furthermore, the above-conceived autocatalytic nuclear runaway process, which is triggered by fuel melting, could occur in any fuel melting situation in which steam bubbles are initially

* A steam explosion at this point is not considered likely, according to the Rasmussen Report, since the coolant is boiling (Ras. Rpt., app. 8, p. B-13). For this reason the coolant channel box is assumed to stay intact to produce the Roman-candle effect. If an explosion occurred, the steam would presumably be vented sideways as well, which might strongly reduce the reactivity momentarily, but only until the steam bubble formed by the explosion collapses or is swept out of the core. Thus, there are other, perhaps even more dangerous courses the accident might follow.

† The rapid interaction of molten fuel and water coolant, forced by the severe power excursion, would more likely generate a steam explosion.

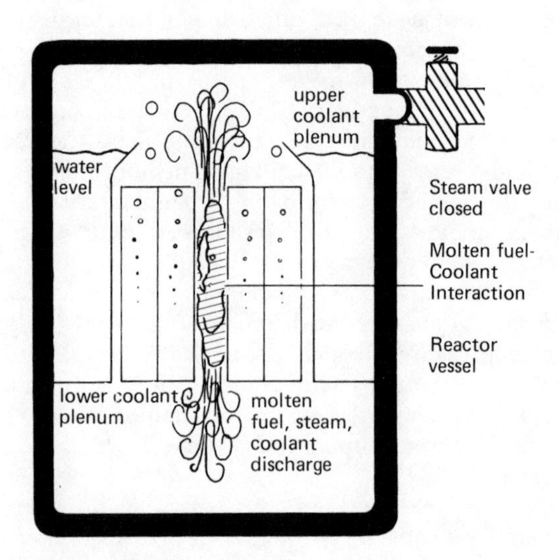

Figure 9. "Roman candle" effect in BWR steam valve closure accident. Note: The core is composed of a large number of long, open-ended metal boxes packed together. Inside each box, or "channel," is a fuel rod bundle. The boxes in which the fuel melting first occurs could constrain the fuel to discharge out the ends, as shown.

present in the BWR core—that is, eliminating the steam valve closure accident would not eliminate this kind of autocatalysis.

It should be noted that the only theoretical analysis of the PEA potential of present-day, large reactors (water-cooled, UO_2-fueled) is contained in an "internal report" (PTR-738) of the AEC's National Reactor Testing Station, which was prepared in 1964. This report calculates a "catastrophic" explosion potential; and is fairly thorough in examining the uncertainties. It also recommends a fundamentally important set of experiments, including full-scale reactor destructive tests to investigate the PEA potential, which we shall discuss later.

Power-Cooling Mismatch Accidents

The most worrisome PCMA involves excessive fissioning or undercooling in a small region of the core while the reactor is operated at full power. Such a situation could be caused, for example, by a misarrangement of the criticality positions of the various control rods, while maintaining steady, full power. Mispositioning of control rods can adversely affect the spatial distribution of the fissioning in the core, causing excessive nonuniformity (local hot spots). Another possible cause for a PCMA would be a blockage of coolant flow to a fuel bundle in a BWR by some foreign object caught up under the core. Such mishaps could lead to excessive boiling of coolant on the surfaces of the affected fuel rods until a thin steam layer or "steam blanket" forms on the surfaces to insulate the rods from the coolant, which will drastically reduce the heat removal from the fuel rods. (This steam-blanketing effect explains why beads of water roll around on a hot skillet or flat iron, instead of boiling off. The beads of water are insulated from the hot surface by a thin steam layer which forms under the beads, so that they evaporate very slowly.) The affected fuel rods could then heat up and melt or crumble in several seconds. The outpouring of molten or crumbled fuel would contact adjacent fuel rods, possibly causing them to similarly overheat by triggering steam blanketing there or blocking coolant flow. The result could conceivably be a self-propagating cascade of fuel rod melting.[18] The concern is that the core could deteriorate within a minute or so to the point where an uncontrollable core meltdown process would be initiated, even if the situation were detected and the fissioning were stopped by a SCRAM. Indeed, initial steam explosions or fuel distortion might jam the control rods and prevent a SCRAM. Then, if a large steam explosion occurred, the effect could conceivably be to produce a severe power excursion by compressing core steam bubbles in a BWR or by blowing off (ejecting) several control rods in a PWR, which would greatly worsen the accident.

Therefore, PCMAs could rupture or explode the reactor vessel with an instantaneous release of the high-pressure, high-temperature coolant and cause a prompt containment rupture due to missiles and pressure waves of steam explosions, followed by whole core meltdown and, finally, by a large fission product release directly into the atmosphere. This accident, like a severe PEA, would be a more severe chain of events than a DB-LOCA without effective ECCS.

Other PCMA possibilities include (*a*) a withdrawal of a single control rod at the normal motor speed, which is too slow to cause a power excursion but which can cause the core power level to rise in addition to increasing the nonuniformity of the fissioning, namely, the intensity of local hot spots in the core (a SCRAM would be ineffective in this situation); (*b*) a withdrawal of a group of rods without SCRAM in a PWR; and (c) a fuel bundle mispositioned in the core (PWR). For these and the previous examples, the core power level would not be reduced, which maximizes the danger of cascading fuel meltdown when steam blanketing occurs.

There are other PCMAs which involve a reactor coolant system fault that would reduce the heat removal from the core but, through negative reactivity feedback, would reduce the core power level as well. Yet, steam blanketing could still occur. An example is an instantaneous coolant pump stoppage (seizure) that would reduce the coolant flow. The coolant in the core would heat up so as to reduce the reactivity and thus reduce the reactor power; but the reduced flow would cause steam blanketing before the core heat (power) output could decline and thus would threaten fuel rod breakup, even if a SCRAM occurred to shut off the fissioning altogether. If a SCRAM failed to occur, this pump-seizure accident would be more severe, since the core power level would remain substantial. In all, there are a great many PCMA possibilities.[19] As with PEAs, full-scale tests would be necessary to verify the theory for those PCMAs in which the reactivity could conceivably rise.

The PCMAs discussed so far involve a failure of the coolant to remove the heat from the fuel. There is also a class of PCMAs not leading immediately to fuel overheating, but leading instead to coolant overpressure, which would threaten to cause a catastrophic reactor vessel rupture or a severe LOCA. These PCMAs involve a failure to remove heat from the coolant.

The coolant's function, besides slowing down neutrons, is to absorb the fission- and after-heat of the reactor core. This heat must then be removed from the coolant before it recirculates through the reactor, or else it will overheat and over pressurize. Normally, the coolant in a PWR gives up its heat by passing through the steam generator. In the process, the boiler water is heated into steam, which is then piped to the turbine. That is, the reactor core heat, which was absorbed by the coolant, is transferred to the boiler water, where it ends up in the steam. The resultant loss of boiler water must be continually replenished, however, and this is done by pumping "feedwater" into the steam generator. To conserve water, the steam, after passing through the turbine, is recycled by condensing it back into water to become the feedwater. But for this to be possible, the heat of the steam must be removed, which is done by the turbine, in generating electricity, and by the "condenser." The condenser cools and condenses the used steam by means of an array of cold tubes through which "cooling water" (for example, from a lake) is circulated. The reactor heat energy, then, ultimately leaves the plant in the form of heated lake water and electricity. In a BWR, the process is the same, except that there is no intermediate steam generator.

Hence, there is a chain of heat exchange processes that must stay intact if the coolant is to be kept from overheating. If this chain is broken—for example, by a feedwater pump failure or by a failure of the outside cooling water pumps or by a closure of the turbine steam valve—then the coolant would heat up and overpressurize. If the pressure surge is not too great, reactor coolant pressure relief valves (safety valves) could vent the overheated reactor coolant into the containment in the form of steam; but in such an event the vented coolant must be condensed, cooled, and pumped back into the reactor (recirculated) to avoid a loss-of-coolant situation in the reactor, which would lead to a core meltdown. (The ECCS could serve to pump water back into the reactor.) The heat of the vented coolant must also be removed by the containment cooling systems to avoid overpressurizing the containment, for a containment rupture could allow the escape of the coolant and thereby lead to a loss of coolant in the reactor. This relief valve-containment cooling scheme, therefore, constitutes another vital chain of heat exchanges. At least one of the chains must work to avoid a serious accident, since, even if the reactor is shut down, the core afterheat, which is substantial, must also be removed and transferred to the outside environment by a heat-exchange chain.

Serious accidents resulting from failures of either or both heat-exchange chains will be referred to as "heat-exchange accidents," to distinguish them from the more direct PCMAs discussed earlier. Specific examples of heat-exchange accidents are as follows:

Turbine valve closing without SCRAM (PWR). This accident stops the flow of steam (heat) out of the steam generator. Since the core remains at a high power level, the reactor coolant quickly overheats and pressurizes up to about 3,000 psi pressure within two minutes,[20] which violates the reactor design pressure of 2,485 psi.[21] Since the reactor design pressure is exceeded, a reactor vessel rupture might be assumed, though there may be enough safety margin to withstand the extra stress. (Recall that this accident in a BWR causes a power excursion, due to the compression of the steam bubbles in the core caused by the pressure surge. This power excursion problem does not exist in a PWR, at any rate, since no steam bubbles are present in the core.) If the reactor system does not rupture, the relief valves would finally take control to reduce the pressure; though other systems would still have to function to shut down the reactor.

Loss of feedwater without SCRAM (PWR). This accident results in a slightly more severe heat-up of reactor coolant than the turbine valve closure case. It could be caused by the failure of the steam condenser, for example. The peak pressure is calculated at 3,016 psi.[22]

Loss of electrical power. Since electric power is needed to run the various pumps in the heat-exchange chains, a complete electric failure would lead eventually to a loss-of-coolant accident, as the reactor would either rupture by coolant overpressure or simply lose coolant.

There are many more heat exchange accident possibilities than would appear from the above discussion, since only a simplified description of the heat-exchange chains is given here. Others are analyzed in the Rasmussen Report under the heading of "transients," some of which could end in a major core meltdown accident and a heavy fission product release.[23] There are so many such possibilities that there needs to be a more tractable way of assessing their likelihood. This problem will be addressed in chapter 6's review of the Rasmussen Report, which may be underpredicting the consequences of such accidents.

Loss-of-Coolant Accidents

The more severe kinds of loss-of-coolant accidents (LOCAs) have only recently been treated in the published literature of the nuclear community with the issuance of the Rasmussen Report. Such accidents involve a failure of the emergency core cooling system (ECCS) or the SCRAM system. While either can produce a core meltdown, the possibility of a SCRAM failure—the failure to insert the control rods following a pipe rupture—needs more attention:

1. For a *large* pipe break in a PWR, a SCRAM *failure* may not make any difference, since the rapid depressurization of the coolant during the blowdown will cause it to flash to steam and thus form steam bubbles in the core (froth), which will ultimately reduce the reactivity and, hence, the core power level to the afterheat level, just as a SCRAM would. Moreover, for the PWR, the emergency coolant is extra borated, so that refilling the core with the borated coolant will not cause a recriticality. (This is because boron absorbs neutrons; see p. 12.)

In a BWR, on the other hand, the emergency coolant is *not* borated, so that refilling the core *would* cause a recriticality[24] and thereby greatly worsen the LOCA. Indeed, in the BWR case there is a question as to whether a power excursion could be triggered *during the blowdown* for the case of a rupture in the recirculation piping. This is because the coolant in such a case would not rapidly depressurize (froth) at first, due to the tortuous coolant flow paths in the BWR (compare fig. 2 with fig. 1). Furthermore, these flow paths are such that for the assumed pipe rupture (recirculation piping) the path of coolant blowdown would be *up* through the core and then down the space outside of the core barrel to reach the ruptured piping. This upward core flow creates a tendency for the bubbles to be swept out of the core during the coolant blowdown by the discharge of the original, bubble-free coolant from below the core, which could conceivably *raise* the reactivity dangerously to trigger a power excursion.[25]

The Rasmussen Report concludes that the power level will remain "significant" during both the blowdown and the emergency coolant phases of the LOCA, making it necessary to assume that a core meltdown will occur, despite the functioning of the ECCS.[26] But the matter cannot rest there; for there is the question of the *severity* of the meltdown. Will it be no more severe than a LOCA without effective emergency cooling (slow meltdown)? Or will a severe power excursion be triggered, which would increase the chances for greater fission product release? To predict the course of a BWR LOCA without SCRAM would be a formidable theoretical undertaking, due to the complexity of calculating the flow of neutrons, the production and losses of neutrons, and the interrelationship these processes have with the fuel and coolant heating and flow processes, which are difficult to calculate in themselves, especially when molten fuel forms and becomes mobile and causes violent coolant boiling and coolant and fuel motion. The net effect of these interrelated processes determines the *reactivity*. Moreover, full-scale reactor experiments would be needed to verify any theoretical prediction, because reactivity increases would be involved.

It is noted that the Rasmussen Report did not reference any theoretical analyses for this LOCA-without-SCRAM situation; and probably none exists; at least, no analysis has been published. Consequently, this author attempted a crude computer simulation of this accident—but just for the coolant blowdown phase. The calculation predicts that no power excursion occurs; though the power level remained significantly high for a period, as the Rasmussen Report contends. Yet, the calculated reactivity does indeed increase and approach recriticality for a period of time, which is disturbing. Further, the theoretical model used is crude and cannot be relied on to rule out the occurrence of a power excursion during the blowdown (the calculation can only show that there are no readily obvious power excursions). For example, one of the ECCS systems will spray cold water into the coolant chamber above the core during the blowdown. This will quench steam and thereby reduce the pressure above the core, which could conceivably accelerate the coolant flow through the core, sweeping out the steam bubbles and raising the reactivity. This effect was not included in the author's calculation, nor was the emergency coolant refill period investigated.

2. It turns out that a small coolant pipe rupture (0.5-square-foot openings or less) in a PWR is serious with respect to a SCRAM failure—that is, it matters whether a SCRAM occurs, which may seem paradoxical to the layman, since a SCRAM is not expected to be necessary for a large pipe rupture (9-square-foot openings). This is because, in a small rupture, a rapid coolant depressurization does not occur to reduce the reactivity; and hence the power level remains high for a period of time. Further, if the rupture occurs in an inlet coolant pipe, the coolant flow in the core would tend to reverse its direction (unlike the BWR). When it does, its momentary stagnation might cause steam blanketing on the fuel rod surfaces,[27] drastically reducing the removal of heat from the fuel rod (see p. 31). The result is that the accident is similar to a PMCA: reduced heat removal from the fuel and a high continuing power level, which could lead to fuel melting in a basically intact reactor vessel containing high-pressure, high-temperature coolant. (The coolant pressure stays above one half of the operating pressure for 100 to 200 seconds, unlike a large-break LOCA in which the coolant pressure reduces to zero in 18 seconds.[28]) The molten fuel then pouring into the hot, pressurized coolant from the melted fuel rods might cause a steam explosion and rupture of the reactor vessel. Missiles (fragments of the vessel), propelled by the sudden release and flashing of the hot, pressurized coolant, might then *promptly* rupture the containment, which is theoretically possible (see pp. 19, 20, 22).

It is possible that the small PWR LOCA without SCRAM could follow a less hazardous but still serious course. For example, the overheated core could crumble into an uncoolable heap, which would then heat up and melt down, after the reactor coolant system depressurizes, despite the ECCS. This could lead to *delayed* containment failure—failure *after* a whole core meltdown occurs —and thus would be essentially the same as a LOCA-without-ECCS type of core meltdown. (Recall that delayed failure of the containment promotes the retention of fission products and therefore a lesser fission product release to the atmosphere than a prompt failure, see pp. 20–23.)

The Rasmussen Report addresses the small LOCA without SCRAM in a PWR; but it only notes that a "core melt" will occur.[29] The report does not elaborate to address the possibility of prompt containment failure, which would maximize the release of fission products; nor does it reference any analysis. Rather, it merely assumes delayed containment failure. Therefore, one must presume that prompt containment failure is possible for this type of accident.

It might be argued that for this small-break LOCA a steam explosion will not occur when the molten fuel pours into the hot, pressurized coolant, as this author assumed in order to postulate the rupture of the pressurized reactor vessel and thus arrive at the prompt containment failure possibility. The ground for this argument is that experiments have been performed with molten metals (not UO_2 fuel) that failed to generate steam explosions with "saturated" water (water which is on the verge of boiling),[30] which might be the condition of the coolant by the time the molten fuel forms in this accident. By that time the loss of coolant will have caused the coolant pressure to drop to near the 1,300 psi "saturation pressure" level at which the hot, 580°F coolant would begin to boil. Higher coolant pressures, such as the normal 2,100 psi operating pressure for the PWR, prevent boiling. (In this respect, this accident differs from a PCMA in a PWR, for fuel melting in a PCMA would occur while the coolant is at full operating pressure, which is far from "saturation".[31] Hence, steam explosions and prompt containment failure are more likely in a PCMA, which is why the PCMA is more serious.) But, as admitted in the Rasmussen Report, no experiments with significant quantities of molten UO_2 have been performed (in either saturated or unsaturated water), and furthermore, the melting temperature of "UO_2 is so much higher than that of the other materials investigated that entirely different phenomena could govern the process. The quantity of molten material that could interact with water in a meltdown accident is orders of magnitude (tens of times) greater than has been investigated in any controlled experiments; such an extrapolation in size would be fraught with uncertainties even if the phenomenology were understood."[32] Because of this, the Rasmussen Report allows the possibility for steam explosions in even saturated water and recommends "large scale" experiments.[33]

One could perform a laboratory experiment in which large quantities of molten UO_2 were dumped into a pool of saturated water; but to duplicate the high coolant pressures would be quite expensive—a large furnace inside a large pressure vessel. Even then, the configuration of melting fuel rod bundles would not be reproduced. Therefore, it may be that the severity of the *small* LOCA without SCRAM can only be established by large- or full-scale PWR tests in view of the uncertainty presented by the many factors involved.

Full-scale tests may be needed in any event to produce the *reactivity effects.* In this regard, we should mention the small LOCA without SCRAM in which the pipe rupture occurs in a reactor *outlet* pipe, instead of an inlet pipe. In this outlet-pipe rupture situation, the coolant flow in the core would not reverse but accelerate, moving colder, inlet water through the core with the possible potential for causing a power excursion.[34] (This question applies to large LOCAs, too, where the rupture occurs in the outlet pipe.) Here, full-scale tests would be needed.

It is concluded, therefore, that the *small LOCA without SCRAM* may be more severe than a *large LOCA without ECCS* relative to fission product release, because of the possibility of prompt containment failure. This conclusion is at odds with the Rasmussen Report, which by assumption places the two accidents in the same fission product release category. This concern is compounded by the possibility that the *small LOCA without SCRAM may be much more likely to occur.* First, a small pipe break is certainly much more likely than the complete shearing off of a main coolant pipe (large break). Second, while there are back-up ECCS systems, there is no back-up SCRAM system (and SCRAM systems have been found totally inoperative, which will be discussed later). Moreover, of the fifty or so control rods installed in a PWR, practically all of them would have to be rapidly inserted in a SCRAM in order to stop the fissioning and avoid core melting, for if as few as two or more control rods fail to insert in a SCRAM, the reactivity evidently could not be sufficiently reduced by the SCRAM to avert core melting.[35] The likelihood of this situation is underscored by the recent discovery of cracks in a certain crucial component within the individual control rod drive mechanisms of several BWRs. If the component were to fail, the affected control rod could not be SCRAMMED. In one instance, twenty-four of the sixty-five drive mechanisms inspected (36%) had cracks.[36] Though this specific cracking situation does not apply to PWRs (wholly different designs), the incidents do show up the possibility of many control rods failing to insert on a SCRAM due to component malfunction. (Indeed, this partial SCRAM failure possibility applies to all reactor malfunctions requiring SCRAM; that is, they all must be evaluated for partial SCRAM failure.)

3. The *small* LOCA without SCRAM in a *BWR* is similarly serious, in that core melting will occur; only the process is different. Specifically, the steam valve is set to trip shut automatically upon a small LOCA to conserve coolant. If a SCRAM failure occurs, then the course of the accident will be much the same as the steam valve closure accident discussed earlier, with a conceivable autocatalytic power excursion (pp. 28–29). Again, the Rasmussen Report notes that core melting would occur but does not elaborate or reference any analyses.[37]

Spontaneous Reactor Vessel Rupture

The consequences of this extreme LOCA were discussed earlier (pp. 20–21). It could be caused by two types of metal failure: (1) cracks developing in the vessel walls due to manufacturing and/or design defects, or maloperation; and (2) cracks developing in the large bolts or "stud bolts" that hold the lid or "closure head" of the reactor down against the high coolant pressure. It is expected that if cracks develop in the vessel wall, they will grow slowly, allowing time to detect the incipient vessel rupture by coolant leakage. The author is not in a position to comment on the reliability of such forewarning. Deteriorating stud bolts, on the other hand, would give no such warning.

The stud bolts are heavily stressed or loaded during reactor operation. Each stud holds back about one million pounds of force.[38] There are about fifty-four closure studs in a PWR. If cracks should develop and one stud fail, its load would have to be taken up by the remaining stud bolts, adding to the stress on each bolt. If other studs were similarly faulty, the process conceivably could quickly deteriorate in a rapid cascade of stud failure, blowing off the closure head, which could pierce the containment. The fission product release in the subsequent core meltdown would then have a direct access to the atmosphere. It should be noted that in three unspecified reactors "two or three" stud bolts "were badly cracked or broken." [39] Incidentally, other accident possibilities exist that involve combinations of a defective closure stud condition and a PEA or PCMA in which a core explosion occurs. In such accidents, the defective closure studs would make the blow-off of the closure head more likely.

In the preceding discussions of PEAs, PCMAs, LOCAs, and spontaneous reactor vessel ruptures, the author has suggested a variety of "conceivable" autocatalytic processes that could lead to disastrous accidents. In several instances, the author has explored these possibilities. However, it must be kept in mind that these conceived processes are but attempts to estimate the severity of possible accidents in the absence of rigorous theoretical analysis, whose development would require large teams of scientists and mathematicians. At the very least, we can say that these estimates cannot be ruled out as impossible; indeed, they are very plausible.

Four

Design Basis Accidents *v.* Worse Possible Accidents

THE SAFETY Analysis Reports (SARs) submitted by the nuclear industry and approved by the AEC (now the Nuclear Regulatory Commission) for licenses to build and operate the reactors do *not* analyze and calculate the consequences of all of the severe reactor accident possibilities. Instead, the SARs consider only very selected accidents that are predicted to result in no core explosion or fuel meltdown or crumbling. These selected accidents are characterized basically by a *single* reactor component or system failure that triggers the accident, such as a single control rod ejection (PWR) or a single control rod dropout (BWR), both being power excursion accidents (PEAs), or a single coolant pipe rupture while at full reactor power (LOCA). The SCRAM and ECCS systems are assumed to function to control these mishaps. The power-cooling mismatch accidents (PCMAs) that are analyzed in the SARs are limited to such PCMAs as a slow withdrawal of a group of control rods without SCRAM in a PWR and a coolant pump seizure with SCRAM in a PWR.

These single-failure accidents serve as the basis for the design of the entire reactor plant and, hence, are called the design basis accidents (DBAs)or DB-PEAs, DB-LOCAs, and DB-PCMAs. In other words, the reactor plants are *not* designed to avert disaster should a *worse possible accident* (WPA) occur involving multiple failures. Some WPAs were mentioned in the previous chapter; table 2 lists these and other examples of WPAs in comparison with DBAs.

Table 2 Some Design Basis Accidents and Worse Possible Accidents

Design basis accidents	Worse possible accidents
Power excursions	*Power excursions*
Single control rod dropout (BWR)*	Single control rod ejection (BWR)
Single control rod ejection (PWR)*	Single control rod ejection (PWR; high reactivity worth)
Steam valve closure without SCRAM (BWR). The coolant pumps are assumed to stop (coast down) automatically in this event, which heats the coolant to reduce reactivity and limit the excursion	Single control rod dropout (BWR),* with high reactivity worth of the control rod
	Single control rod dropout without SCRAM (BWR)
	Single control rod ejection without SCRAM (PWR)
	Cascading control rod ejection
	Steam valve closure without SCRAM, and without automatic coolant pump stoppage to reduce reactivity (BWR)
	Cold water reactivity accident (PWR)
	Core dropout (PWR)
	PEA with defective reactor vessel closure bolts
Loss of Coolant	*Loss of Coolant*
Largest coolant pipe rupture at steady, full power*	LOCA with failure of the ECCS
	Pipe rupture due to a coolant pressure surge following a core overheating event (PEA or PCMA)
	Spontaneous reactor vessel rupture
	Small pipe rupture without SCRAM

Design basis accidents	Worse possible accidents
Power-cooling mismatch	*Power-cooling mismatch*
Coolant pump seizure (PWR)*	Control rod position misarrangement
Continuous withdrawal of a group of control rods at normal speed and low reactivity worth of the control rods without SCRAM (PWR)	Flow blockage to one or more fuel bundles (BWR) without SCRAM
Flow blockage to a fuel bundle (BWR)	Coolant pump seizure without SCRAM (PWR)
	Continuous withdrawal of a single control rod with or without SCRAM (PWR)
	Continuous slow withdrawal of a group of control rods with high reactivity worth and without SCRAM

* With SCRAM

There are a great many WPA possibilities, which the AEC and the nuclear industry dismiss essentially on the basis of subjective judgment that the chance that any WPA will occur is remote because of certain elaborate measures taken to avoid them. Hence, the WPAs, and the possibility for accompanying autocatalysis, are completely neglected in the SARs and in the AEC's reactor safety research program; yet, practically all of the reactor accidents that have occurred in the past have been multiple-failure accidents; [1] so any claim that a disastrous WPA is not likely ever to occur is only conjecture, despite sincere human attempts to prevent accidents and to estimate their probability.

Incidentally, the past reactor accidents have occurred in very small, early reactors and involved little radioactivity. The reactors used metallic fuel, which, because of its low melting temperature, releases a minimum of fission products upon melting; and, furthermore, there was no high-temperature, high-pressure coolant to flash into steam to aid in expelling what radioactivity was released from the core. Finally, the reactors were in remote locations, except for the Fermi partial fuel meltdown accident (which is discussed in app. 2). These factors explain why the consequences of the accidents in the early reactors were not disastrous.[2]

The Rasmussen Report addresses some of the WPAs, as we shall see when the report is examined in chapter 6. However, they are generally the less severe ones (though several of them are very serious), and they are given only cursory treatment in regard to the course they each could take. Moreover, one is asked to accept the report's conclusions on faith, as no theoretical analyses are given or referenced with which one could systematically check the results, assumptions, and theory. (In the case of the DBAs, at least, some review can be undertaken through the SARs and the technical reports referenced therein.) There is one exception in the Rasmussen Report: a LOCA without ECCS, which is treated in some detail; but this LOCA, we will recall, is the least severe class of serious accidents. Most of the Rasmussen Report consists of attempts to estimate the probability of the selected WPAs.

Having explored the full accident potential of water-cooled reactors, we turn now to the Design Basis Accidents and evaluate the *experimental basis* for the conclusions of the nuclear community that the DBAs will be controlled safely should they occur. It will be shown that the DBAs have not been demonstrated scientifically to be safely controllable. This is perhaps of more immediate concern than the WPAs, since the DBAs could be considered more likely to occur. (This evaluation will also serve to introduce the reader to the experimental problems of investigating reactor accident hazards, so that he may better appreciate the experimental requirements and myths surrounding the worse possible accidents, which will be discussed at the end of this chapter.)

Design Basis, Power Excursion Accidents

The reactor physics theory of neutron dynamics, usually called neutron kinetics, used to calculate the "energy yield" of the DB-PEAs— that is, the heat produced by the power excursion— has never been verified by any experimental power excursion in a *large core*. Though some limited power excursion tests were performed about ten years ago—the Special Power Excursion Reactor Tests (SPERT)—these involved extremely small reactors, about 1/500 of the volume size of today's large cores.[3] The most trusted theory of reactor dynamics predicts that large cores will exhibit neutronic effects, called "space-time kinetics effects," which will cause much stronger excursions than have been predicted by small-core theory, called point kinetics.[4] Therefore, verified small-core theory is wholly inadequate for large-core situations; hence, the small-core experiments are of little help. As the AEC's *Water Reactor Safety Program Plan* stated in 1970: "[T]he experimental basis for determining the physical correctness of space-time calculational results [for large cores] does not exist as it does for small cores." [5]

Moreover, the DB-PEA calculational methods are founded on numerous mathematical approximations to the theory, in order to perform practical calculations with a computer. Specifically, the DB-PEA calculations are based on a simplified theory of neutron dynamics known as "diffusion theory," a mathematical approximation of the rigorous "neutron transport theory," the error of which has never been determined by power excursion experiments or DB-PEA calculations with the rigorous theory.[6] In addition, it is not practical to calculate the DB-PEAs by "directly" or strictly applying even the diffusion theory, due to excessive computer demands; and so further approximations are made by the use of "approximate" calculational methods, called indirect methods, for DB-PEA analysis—that is, mathematical approximations to the diffusion theory—and this introduces an additional source of error, error relative to diffusion theory. There are no mathematical bounds of this additional source of error, which compounds the uncertainty in the DB-PEA calculations.[7] "Direct" calculational methods based on diffusion theory are being developed, namely, "finite difference" methods, which are to be used to check the accuracy of the DB-PEA calculations (within the limitations of neutron diffusion theory), but only for several selected accidents, called "benchmarks." [8] However, since these finite difference methods are evidently not practical for calculating more than a few of the DB-PEAs, and since there are no mathematical error bounds for the "indirect" calculations to use in fixing the direct-to-indirect error of the calculation of those DB-PEAs away from the benchmarks, we will not really know the magnitude of this second source of error for the bulk of the DB-PEAs. This latter point may seem minor, but it is an example of how predictions of safety are not well established in regard to accuracy. Also, the direct, finite difference diffusion theory method has not been applied correctly in any benchmark calculation to establish true or accurate calculational results (again, aside from the inherent error of diffusion theory relative to transport theory) for comparison with the indirect diffusion theory, DB-PEA calculational methods; that is, the necessary number of mathematical "mesh points" for DB-PEA conditions has not been determined, as will be discussed shortly. It is important to identify all sources of error and uncertainty, since there is little margin for error in the DB-PEAs. (The PWR DB-PEA calculations are more likely to be conservative than the BWR calculations, due partly to the fact that the reactivity increase potential of the PWR control rods is less than that of the BWR.) The only way to verify the calculational predictions is to perform power excursion tests using *full-scale reactors,* because of the before-discussed peculiarities of the physics of neutron kinetics. No such tests have been run, and none are planned, despite the fact that the AEC's chief reactor test laboratory recommended such tests in 1964 in an "internal report." [9]

As it is, the unverified PEA theory predicts that the peak energy yield per gram of fuel of the DB-PEAs approaches rather closely (by about 70%) the design safety limit of 280 calories/gm[10] (270 cal/gm corresponds to the UO_2 melting temperature).[11] Thus, a 30% error or more in the calculations could mean disaster; and such an error could easily exist, in view of the enormous complexity of the predicted phenomena, the sensitivity of power excursions to reactivity errors, and the lack of any experimental verification whatsoever.

Moreover, in 1969 the power excursion theory in the form of the "industry standard" computer program, called WIGLE,[12] was given a partial test of its simplest aspect, and it "completely failed" in the unsafe direction—a fact which has been left buried in the infinite number of reactor technical books and reports.[13] Specifically, the test consisted of a burst of neutrons injected into the side of a reactor, which was large in one dimension but which was not quite critical; that is, it was not quite capable of sustaining a steady fission reaction rate. Since the core was not critical, the number of neutrons in the core after injection was destined to rapidly diminish to zero while diffusing rapidly across the core as a wave of neutrons. The neutron wave was then measured as it propagated across the core for comparison with the theoretical prediction of WIGLE. This type of neutron transient allowed a clean experimental test of the basic, non-feedback, constant reactivity aspects of the theory, as there were no fuel heating and other core changes occurring that would have complicated the transient and associated theoretical predictions. Yet, the WIGLE DB-PEA theory failed to predict the neutron wave characteristics (mainly, the time duration of the wave) and failed in such a way as to indicate that the theory might underpredict the energy yield of a power excursion by one half. This discrepancy is not factored into the DB-PEA calculations. Since the theory failed to predict a simple neutron transient, one wonders how much more inaccurate the theory might be for power excursions. Especially disturbing is the fact that the recommendations of the National Reactor Testing Station (NRTS) laboratory in 1970 for a special research program to attempt to resolve the discrepancy by, for example, using a more advanced neutron dynamic theory, called TWIGL, were not approved by the AEC.[14]

The difficulty in trying to force WIGLE to agree with the results of the neutron wave experiment may very well be due to the fact that the neutron wave itself violates the mathematical assumption underlying the WIGLE theory. This neutron kinetic theory is based on the simplified diffusion theory of neutron dynamics, which is a mathematical approximation to the more exact "transport theory."[15] The error of this approximation can only be found by comparing with transport theory calculations, and this has not been done. The diffusion theory would be an adequate approximation of the transport theory, however, if one could assume that the neutrons are flying around equally in all directions, or nearly so, but this situation does not hold true for the front and rear of the wave in the aforesaid experiment. Therefore, it may be that transport theory is necessary to predict adequately the neutron wave experimental results; but this author is unaware of any computer program that can use the transport theory for PEAs in water-cooled reactors, for we are up against the limits of computer capability even with the less difficult diffusion theory.[16] Perhaps the discrepancy of the experiment will never be resolvable, due to an inability to calculate the neutron wave processes by transport theory—at least not within, say the next five years.

On the other hand, neutron waves as such are not expected in PEA situations so that the presently used diffusion theory assumption might be adequate for PEA predictions. This is because the process of injecting neutrons in one side of the reactor does not occur in a PEA. Rather, the reactor in a PEA creates its own neutrons throughout the core. There is still some similarity, however, inasmuch as the neutron density level will grow very rapidly and nonuniformly in a PEA, so that transport theory may still be necessary. Even if we could use transport theory in a practical computer calculation, it would have to be in a simplified form, such as approximating the continuous spectrum of neutron speeds by a few speed "groups;" and any simplified (practical) transport theory will contain approximating assumptions which inject other unbounded errors. Since the completely rigorous form of transport theory, which we might trust, could never be used as a basis for calculation to determine the error of approximation, the only way these uncertainties can be resolved is by power excursion experiments in a large reactor, which have not been performed.[17] Such experiments were actually proposed in 1966 by the National Reactor Testing Station laboratory (NRTS) under the title, "Large Core Dynamics Experimental Program."[18] Phase I of this program was the above-discussed neutron wave experiment. Phases II and III were to be the power excursion experiments, specifically, "control-rod-induced power excursions." The entire program was estimated to require five years to complete the first power excursion experiment (1973 completion date) and three more years to complete the rest, ending in 1976.

In justifying the three-phase set of experiments, the NRTS stated in its technical proposal:

> Because of the mathematical complexities encountered in the formulation of the large-core dynamics problem, practical methods of large-core dynamics analysis are necessarily approximate. Before these methods can be applied with confidence in safety analyses, experimental verification of their validity is required. . . .
>
> Independent of the nature of the space-time dynamics [calculational] techniques ultimately developed, they cannot be used as a basis for predicting the courses of the reactivity [power excursion] accident until their accuracy and range of validity have been firmly established. This can only be accomplished by comparision of the analytical results with definitive experimental data. At the present time no such data exist. . . . In view of the increase in the physical dimensions of reactor cores to be utilized in the power reactor industry, the need for such data is urgent.[19]

However, the proposed large-core power excursion experiments (Phases II and III) were never carried out, nor were any similar or substitute experiments ever performed,[20] despite such clear, authoritative opinion that they are necessary.[21]

The fact that the WIGLE power excursion theory failed to be verified by the neutron wave experiments of Phase I of the proposed Large Core Dynamics Experimental Program underscores the need and urgency for the program. The question arises: How could the AEC have justified authorizing the construction of any large reactors (the AEC has authorized over 103 reactors since 1964) in the face of no experimental data to verify the PEA theory and, indeed, in the face of the WIGLE discrepancy? We shall take up this question later.

It is noteworthy that the NRTS technical proposal was issued in 1966 as an "internal report" (PTR-815) and was never revealed to the public, nor was it cited in the open literature, until the NRC released the document in December 1975.[22] In addition to this document, there apparently exists a separate, draft-plan document for the Phase II experiments.[23] The NRC should make this document available to the public as well.

Meanwhile, to revise the present DB-PEA predictions to at least provide a safety factor that allows for the WIGLE discrepancy, neglecting the scientific uncertainty of the DB-PEA predictions due to the lack of the Phase II and III data, would mean that present water-cooled reactors (at least, the BWR) could not meet the DBA safety criteria, and would probably require a major redesign of water-cooled reactors to reduce the reactivity potential of the DB-PEAs. This redesign effort would be impractical or, at any rate, would take ten years or so to accomplish.

However, there is the possibility that the WIGLE discrepancy can still be resolved, though this would lead to other difficulties, as we shall see next.

Professor A. Henry of MIT has recently revealed that a reanalysis of the above-discussed neutron wave experiment has just been performed (to be made public), which allegedly has resolved the WIGLE discrepancy.[24] The discrepancy is now being attributed to the fact that the WIGLE theory calculates the neutron wave process in only one dimension of a reactor, therefore neglecting the effects of the other two dimensions. Specifically, a two-dimensional diffusion theory calculation, like TWIGLE, which was mentioned earlier as an improvement on WIGLE, has now been carried out which "predicts" the experimental neutron wave data accurately, according to P. Parks of the Savannah River Laboratory.[25] It is reasoned that a two-dimensional calculation is needed because the neutron process in the experiment was complicated by the presence of a "neutron reflector," placed partially around the reactor in a way that introduced a second-dimensional effect. In power reactors, a neutron reflector exists completely around the reactor core, which, by Parks' finding, would require a three-dimensional calculation for power excursion safety analysis.

What could we conclude from this tentative reanalysis? For one thing, the WIGLE theory is not reliable for safety analysis purposes. Since the nuclear industry's BWR design basis, power excursion accidents are checked with WIGLE, these DB-PEAs must now be recalculated using a three-dimensional calculation. Inasmuch as WIGLE erred in the unsafe direction, the three-dimensional calculations may predict worse power excursions.

However, it has not been shown that it would be practical to recalculate the DB-PEAs using a three-dimensional calculation, for in order to obtain sufficient accuracy with a 3-D calculation, the reactor may have to be mathematically subdivided into a fine three-dimensional "mesh." (In principle, the computer calculation would be exact within the limits of the diffusion theory, if the subdivision were infinitely small.) But this could make the computer calculation impractical, by requiring extremely long computer running times to generate answers. In the above-mentioned NRTS report (PTR-815), it is asserted that about one million mesh segments or "mesh points" are needed "if accurate results are to be obtained." [26] Implying that "several hundred thousand spatial mesh points" are needed to solve a "large [reactor] three dimensional problem," Professor Henry, who is a foremost expert on such matters, stated in 1970 that a 3-D calculation "still seems to be beyond the capacity of the present generation of computers." [27]

For comparison, MIT has developed a three-dimensional calculational method; but it has only been used with 6,500 mesh points.[28] Similarly, only 1,989 mesh points were used for PWR power excursion calculations.[29] Neither number has been shown to be sufficiently accurate for safety analysis purposes by comparison with finer-mesh calculations.

The reanalysis of the neutron wave experiment does not rest on these observations alone, however. For it must be noted that the "success" of the two-dimensional calculation by Parks to "predict" accurately the neutron wave phenomenon depended on using the neutron transient data obtained from the experimental results as input data for the calculation, making the calculation not a true prediction. Specifically, the calculated *reactivity* of Parks' two-dimensional theory had to be adjusted to fit the value that was measured by the neutron wave experiment, in order to avoid gross error. (This needed to be done in the WIGLE case also.[30]) This further supports the need for large-core power excursion experiments, since in a power excursion accident we would not know beforehand how the reactivity, including feedback, will vary with time. The reactivity, after all, is the primary quantity that must be calculated by the theory, and this must be done very accurately. Indeed, both Diaz and Ohanian and the National Reactor Testing Station, who conducted the experiment, pointed out this shortcoming and further concluded that power excursion experiments are the "next natural step" that must be taken.[31]

Nor is it clear that the "WIGLE discrepancy" has been resolved by Parks. The claim is preliminary in nature, and remains to be published and scrutinized.[32] For one thing the mathematical mesh was coarse and was not made fine to check on the size of the numerical error. Moreover, there is still the troubling matter that the neutron wave would seem to require neutron transport theory over diffusion theory.[33] In this regard, it should be noted that reactor control rods add transport theory effects in the core. Yet, the reactor in the neutron wave experiment was not interspersed with control rods, as is the case in nuclear power reactors. It may be that the experiment was too "clean" a test of the theory to be of any practical use. Indeed, a single row of control rods was used in the experiment to examine such effects, and the WIGLE discrepancy became worse.[34]

Also, according to Diaz and Ohanian, "the presence of two-dimensional effects was ruled out" by other dynamic measurements, which seems to contradict the tentative two-dimensional hypothesis of Parks.[35] Diaz and Ohanian at least recommended additional experiments, such as a core with no neutron reflectors, to isolate the cause of the discrepancy; but these recommendations were never adopted.[36] Thus, it might be that the alleged success of the two-dimensional calculation by Parks was fortuitous—at any rate, the matter isn't sufficiently understood.

Regardless of the predictability of neutron wave experiments, there remains the need to perform large-core power excursion experiments to check the accuracy of time-varying neutron diffusion theory in DB-PEA situations, when reactivity changes will be present—namely, the large, nonuniform changes in the composition of the core that change (raise) the reactivity initially and the reactivity feedback effects (due, for example, to the rapid fission heating). These two fundamental aspects are not present in a neutron wave experiment. In order to appreciate the need to verify predictions of time-varying reactivity changes, we should examine our experience with those reactor phenomena which have been observed, involving three-dimensional neutronic effects.

One such instance was the calculated prediction of the core excess reactivity during the gradual burn-up of the fuel over the long-term normal power operation of the Shippingport pressurized water reactor. This is a time-varying process with feedback, since the fuel and control rod conditions change as the fissioning occurs. The Shippingport core also included special nonfissioning, neutron-absorbing material besides the control rods to help limit the excess reactivity, similar to the use of boron in the coolant of commercial PWRs. As the fuel was to be consumed, the special absorber material was designed to be consumed also, so that the gradual negative reactivity feedback of fuel consumption would be offset by the positive reactivity feedback of the depletion of the consumable absorber. The consumable absorber thus slows down the depletion of the excess reactivity and allows a greater amount of fissionable fuel to be placed in the core for longer reactor operation between refuelings. Actually, the excess reactivity was designed to rise from its beginning-of-life value for a period before eventually declining to zero, when the core must be refueled. The Shippingport fuel depletion process, therefore, is a time-varying process with complicated reactivity feedback effects. A space-time neutronic "synthesis" calculation for approximating 3-D neutron diffusion theory was used to predict the fuel depletion for this Shippingport PWR depletion problem.[37] The synthesis method used is similar to the crude approximation methods now used for reactor power excursion calculations. Prior to the operation of the Shippingport reactor, the diffusion theory had been adjusted to agree with the neutronic characteristics of the core as measured during start-up tests; so that subsequent fuel depletion was to be a test of the ability of the diffusion theory approximation to predict the feedback effects of fuel depletion. As the fuel was depleted during reactor operation, it turned out that the excess reactivity actually rose to a value which exceeded the prediction by .9% (reactivity units), which is very substantial. When the discrepancy first appeared, the Shippingport designers reported the following:

> The implications of this data is two-fold: (1) the reactivity level of the core is about 0.5% more reactive at this time than estimated from the design depletion calculations and (2) the increase reactivity condition of the core relative to the calculational estimate (at this [early] core age) indicates a possible calculational misrepresentation of some aspect of the three-dimensional nuclear depletion characteristics of the reactor. [WAPD-MRP-113, p. 85]

Later, when the discrepancy grew larger, it was reported: "Based upon the above results, it appears that Core 2 is about 0.9% more reactive, at power, than the design calculations would have predicted" (WAPD-MRP-114, p. 96). If a reactivity feedback error of such magnitude were to occur in a DB-PEA, the consequences could be disastrous. Therefore, this Shippingport experience, though not applicable to power excursions, does underscore the possibility of errors being present in unverified theory and thus the need for power excursion experiments.

The foregoing discussion shows that the federal government's reactor safety research program has not taken care to rigorously establish safety margins with respect to the DBAs, an important consideration in view of the little margin for error, at least for DB-PEAs. It has been six years since the final report of the neutron wave experiment; and the nuclear community is just now coming around to seeking a resolution of the experimental-WIGLE discrepancy. Meanwhile, about fifty-five large power reactors have been approved for construction.

We turn now to the question of fuel rod integrity in a DB-PEA. The 280 cal/gm excursion design limit, which is intended to assure that the fuel rods stay basically intact in a DB-PEA (see p. 46), has not been demonstrated by valid experimentation. The concern here is not that the reactor might promptly explode at fuel energy bursts less than 280 cal/gm but that the fuel might disintegrate (crumble) and thereby lead into an uncoolable fuel situation (see p. 19). The 280 cal/gm value is founded only on excursion testing of fresh, unirradiated (zero "burn-up") fuel rod specimens, when the fuel is strongest. (This testing is conducted using a small reactor which is designed to generate power excursions with the damage limited to the fuel rod test specimens being tested.) Such fresh fuel testing is *insufficient*, since the radioactive fission product hazard is greatest when the fuel is in a high-level burn-up condition. It is well known that when fuel rods operate in a reactor, they become swollen with burn-up, corroded, and embrittled by irradiation—in short, weakened to near the point of cladding failure.[38] *Thus, tests with irradiated fuel rods are necessary.*

In 1969–70, tests were finally performed on "low burn-up" fuel, and fuel failure was found to occur at 147 cal/gm, well below the PEA design limit and the maximum calculated DB-PEA values. In an AEC "limited distribution" test report, which has been kept secret, the NRTS concluded that the "unexpected" results

> bear directly upon LWR [light water reactor*] safety and cast serious questions about the adequacy of existing information and analytical models used in safety analysis, . . . [that the] implications are crucial, . . . [and] that additional test data [on high burn-up fuel] are urgently needed.[39]

In subsequent tests of higher burn-up fuel (65% of maximum designed burn-up), a fuel rod specimen "failed completely" at

* The technical term *light water reactor* means a reactor cooled by ordinary water and includes the PWR and BWR.

85 cal/gm. In the test report the NRTS concluded that the data were "alarming" and that there is a lack of scientific understanding of the behavior of irradiated fuel rods in a DB-PEA.[40] However, the report does not elaborate to discuss the significance—for example, the fuel crumbling question. This second test report was printed, but was never issued even as a "limited distribution" laboratory report.[41] The 85 cal/gm test failure and photographs of the failed fuel rod specimens were reported in an article in *Nuclear News* in 1970, but the significance was not addressed.[42] The article notes only that the experimental program was "limited" and that "further investigation" is warranted.

The troublesome irradiated fuel test data may have been disregarded by the AEC on the ground that the fuel test specimens were not irradiated prior to excursion testing in a way that is characteristic of commercial reactor operations. Specifically, the test rods were caused to undergo relatively frequent changes in power output during irradiation, which might have weakened the fuel cladding. On the other hand, the cladding of the test specimens was not exposed to the proper level of fast-neutron bombardment during the irradiation. Fast neutrons damage the cladding by making it brittle, which suggests even more strongly the possibility for crumbling. Hence, the cladding was not as damaged from this standpoint as it would be in an actual commercial reactor. (The *Nuclear News* article said that the "damage to the cladding" of the specimens by fast neutrons was "insignificant," which is "atypical of BWRs.")[43] Moreover, the test fuel rod specimens contained less fission product gases trapped in the UO_2 fuel material (as a result of the uncharacteristic irradiation process) than is the case for normal, commercial fuel rods. These gases, conceivably, would tend to cause the fuel to distintegrate when heated in a PEA.[44] All in all, the fuel may fail in an even worse manner than indicated by the above-mentioned failure data.

If this failure data is to be disregarded, however, we are left with no data at all for the irradiated fuel condition with which to assess whether the fuel will remain intact in a DB-PEA. Clearly, it would be prudent to heed the test data, until and if future data refute it.

It should also be mentioned that the irradiated fuel test specimens in those tests were single rods; whereas a multi-rod test of a bundle of five fresh fuel rods, surrounded by a canister "to nearly simulate" actual reactor conditions,[45] resulted in *fuel crumbling* at energy yields as low as 225 cal/gm, due to mutual rod-to-rod heating effects, (again these results were given "limited distribution").[46] Since the core is composed of tight bundles of fuel rods (totaling 30,000 rods or more), *irradiated* fuel could break up in an even more dangerous manner in a DB-PEA than the single-rod irradiated fuel tests have indicated. Instead of pursuing the matter, the fuel test program was abruptly terminated.[47] Publicly, the AEC noted only that the safety research programs were being "reoriented" away from excursion testing fuel, allegedly for another reason, that being that the probability of PEAs is judged by the AEC to be low enough to justify deemphasizing such testing.[48]

The Nuclear Regulatory Commission recently rejected the concern for fuel rod crumbling occurring below 280 cal/gm in a DB-PEA. Said W. V. Johnston, the NRC's chief of the Fuel Behavior Branch, Division of Reactor Safety Research:

> Let me first point out that I find no evidence in the literature of any quantity of fuel being expelled from fuel rods, irradiated or unirradiated, at energy depositions of less than 300 cal/g average. The *Nuclear News* article carefully points out (P68) even for the case of the irradiated rod that failed at 95 cal/g that hardly any fuel was expelled from the rod. It therefore seems to me to be unduly pessimistic to assume such a low value for fuel dispersal.[49]

In support of this assertion, the NRC cited three documents which purport to review all of the excursion testing.[50] However, none of these documents mentions the multirod fuel crumbling test results or references the associated test report. The NRC's Johnston later confirmed that he was unaware of the report.[51]

So far, the use of plutonium fuel has not been mentioned. Plutonium is a by-product of reactor operation due to the conversion of some of the uranium into plutonium by certain nuclear processes that occur in the core, and is of value as a fuel resource.[52] Since plutonium is also an atomic bomb material, there is a need to consume it as fuel in a power reactor, rather than letting it accumulate with the growing risk of theft and the consequent uncontrolled spread of atomic bombs.[53] It is planned, therefore, to "recycle" plutonium as fuel in power plants in the form of a mixture of plutonium oxide and uranium oxide (PuO_2-UO_2). Unfortunately, it appears that plutonium fuel rods may be even more susceptible to crumbling than UO_2 rods. This is because the PuO_2 will exist as particles embedded in "depleted" UO_2 material (UO_2 depleted of the naturally occurring uranium 235, the rare fissionable grade of uranium). The heat energy of an excursion, then, would be concentrated in the PuO_2 fuel particles, so that the PuO_2-UO_2 rods would tend to explode below the 280 cal/gm excursion energy level.[54] Indeed, in one test of a fresh, unirradiated PuO_2-UO_2 rod, described in a "limited distribution" report,[55] the specimen broke into pieces at 260 cal/gm or less. No testing of *irradiated* PuO_2-UO_2 fuel rods or of multirod bundles of PuO_2-UO_2 rods has ever been performed; so the failure characteristics of plutonium fuel may be even worse than suspected.

Lower cal/gm fuel failure thresholds also complicate the neutron kinetics (reactivity) calculations by causing complex fuel and coolant motion and associated reactivity feedback effects, which would tend to make DB-PEA calculations even more difficult or impractical and which would greatly increase the number of full-scale DB-PEA tests required to adequately verify a DB-PEA calculational theory for use in a DBA safety analysis. (For example, could fuel rods breaking into pieces and shifting around cause autocatalytic reactivity feedback?) Moreover, a serious under-prediction of the energy yield of a DB-PEA due to errors in the large-core, neutron kinetics theory, and decreased fuel rod failure thresholds, could conceivably combine to produce an autocatalytic response or otherwise effect a disastrous accident, either through instant core meltdown or explosion or through fuel crumbling or cascading fuel rod failure with successive fuel heat-ups, meltdowns, and explosions.[56]

In order to adequately investigate the DB-PEAs, at least fifty full-scale DB-PEA tests would have to be performed to cover the range of various combinations of different factors. These factors are: the two reactor types (BWR and PWR); various types of DB-PEAs for each reactor type; several fuel burn-up levels; various initial power levels and coolant conditions; and other various core conditions that can affect the reactivity trace of a PEA. The practicality of such an experimental program is doubtful. Two full-scale, that is, full-size, commercial-type reactors (a BWR and a PWR) would have to be built deeply underground or with extra strong containment for use as the test reactors. The containment would have to be designed to contain the maximum explosion and pressures that could conceivably arise in such testing, should the DB-PEA safety predictions prove wholly erroneous. It may be that the tests themselves would be too hazardous. Also, about fifty full-size commercial reactors would have to be operated for an average of three to four years just to preirradiate the test cores. This is about the present number of nuclear power plants already built. But one could not simply use the spent fuel elements of operating reactors. Rather, whole cores would have to be removed as they exist in the various particular stages of use being PEA-tested, since the distribution of the fuel composition and burn-up levels would be a primary influence on the course of PEAs and would, therefore, have to be duplicated. This requirement would further complicate the cost of the tests. Moreover, the operation of these fifty reactors for four years involves large risks in itself, inasmuch as the safety of DB-PEAs is not established.

If any explosion, fuel melting, or even incipient fuel crumbling were observed in any of the DB-PEA experiments outlined above, contrary to industry prediction, water-cooled reactors (PWRs and BWRs) could not be judged safe relative to DB-PEAs. For even if the damage were limited to safe levels in a particular test or tests, the complexity of the phenomena, and the impossibility of measuring most of the detailed core behavior during a PEA, would prevent the development of a practical, reliable theory with which to prove that any of the many more combinational possibilities of DB-PEAs not tested would not result in a disastrous explosion. In addition, there is a chance factor of steam explosions, which means that, if no explosion occurred upon fuel melting in one test, the same test repeated might generate one. Resort to additional testing to pursue any such troubling results would definitely be impractical.

Before such an experimental program could ever be started, meaningful "separate effects" testing would have to be performed: for example, resolution of the experimental error of WIGLE; large-core power excursion tests to investigate the validity of the neutron dynamics (kinetics) theory under reactivity growth and feedback conditions; and proper, multirod excursion testing of fuel—irradiated and unirradiated. Unfortunately, no facilities exist for adequately conducting these separate-effects experiments, notwithstanding the Power Burst Facility (PBF).[57] True, the PBF can excursion-test fuel rods; but due to the short length of the PBF core and other PBF drawbacks,[58] and the lack of a credible fuel failure theory, the results of PBF testing could not be reliably extended to actual reactor situations. For example, there is some indication that the PBF may not have adequate capability for multi-fuel-rod testing of moderately high burn-up (irradiated) fuel.[59]

It is, however, conceivable that an alternate experimental approach designed around separate-effects testing might be sufficient. This would require only one full-scale reactor (preferably, a BWR) for testing the neutron kinetics theory by generating power excursions. (The BWR could fairly simulate a PWR by adding boron in the coolant and testing without steam bubbles in the core.) The fuel could be fresh (strong) and be without plutonium, so there would be no large radioactivity hazard, and could be varied in composition to simulate nonuniform fuel burn-up effects, for example. Then, a full-length, larger version of the PBF test reactor would have to be built to ensure that the fuel rods are more properly tested. Used fuel bundles from actual commercial reactors would have to be used as the test specimens. Such a program, which would cost about $1 billion, would be aimed at establishing a safety-limit criterion for ensuring against fuel crumbling and a neutron kinetics theory for predicting whether a given power excursion will yield fuel energy values that stay below the safety limit. This piece-by-piece approach has great appeal, but would still be unsettling, since actual DB-PEAs would not have been tested. Whether this alternate approach will be adequate will require more analysis than now exists—for example, a full, thorough review of the limitations of the PBF.

All things considered, an adequate experimental program for DB-PEAs would take at least ten years and perhaps five or more billion dollars or may very well be impractical. Without such experiments, the DB-PEA predictions must be considered unverified and therefore unreliable.

It is concluded, therefore, that it has *not* been scientifically established that water-cooled reactors (PWRs and BWRs) are fail-safe with respect to single-failure, design basis, power excursion accidents, which is a minimum safety criterion for any responsible technological undertaking. Indeed, the AEC's *General Design Criteria* for nuclear reactors reaffirms this criterion:

> The reactivity control systems shall be designed with appropriate limits on the potential amount and rate of reactivity increase to assure that the effects of postulated reactivity accidents can neither (1) result in damage to the reactor coolant pressure boundary greater than limited local yielding [for example, reactor vessel rupture] nor (2) sufficiently disturb the core, its support structures or other reactor pressure vessel internals to impair significantly the capability to cool the core. These postulated reactivity accidents shall include consideration of rod ejection (unless prevented by positive means), rod dropout, steam line rupture, changes in reactor coolant temperature and pressure, and cold water addition.[60] [The AEC's WASH-1270 report adds the BWR *steam valve shut-off accident* with SCRAM failure to the category of DB-PEAs to be fail-safe.] [61]

However, the AEC (NRC) judges the safety of the DB-PEAs—that is, whether they satisfy this safety design criterion—on the basis of unverified theory and by ignoring alarming experimental results. The matter is extremely serious, in view of the enormous radioactivity hazard involved and in view of the potential for catastrophic, autocatalytic nuclear runaway.

Design Basis, Power-Cooling Mismatch Accidents

As for the PCMAs, no experiments with fuel rod bundles or even single rods have ever been performed,[62] including those concerning the possibility for cascading fuel rod failure triggered by a localized core overheating mishap. Planned testing in the PBF related to limited PCMAs should shed some light on this type accident; but these tests have shortcomings,[63] and the data still remain to be obtained and evaluated.

The nuclear industry's analyses of the DB-PCMAs predict that steam blanketing could or will occur (see p. 31) but that the affected fuel rods (about 10% of the rods in the core, except for the case of coolant flow blockage to one fuel rod bundle in a BWR) will not heat up enough to cause serious core damage—namely, cascading core melting. However, no basis is given for this conclusion, other than a theoretical heat-up prediction that the cladding won't exceed an assumed disintegration temperature of 2,700°F. Nor are the theoretical predictions explained and justified, or referenced,[64] except for the BWR fuel bundle blockage case, which will be discussed shortly. In contrast, the AEC's main reactor test laboratory reviewed the "present capability" to predict the outcome of PCMAs in a report on part of the planned PCMA testing in the PBF and concluded that there is gross uncertainty, due to the complex, interacting processes occurring within a fuel bundle—for example, a 450°F uncertainty in the fuel temperature in the pre-steam-blanketing phase alone.[65] Following steam blanketing, the laboratory concluded, the predictions of fuel rod behavior in PCMAs have a low confidence, mainly because of the "lack of experimental data." [66] In short, the controllability of DB-PCMAs has not been established.

The BWR fuel bundle flow blockage accident (see p. 31) is a case in point. The BWR designers, General Electric Co. (GE), have issued a theoretical analysis of this PCMA,[67] which predicts that steam blanketing and severe fuel rod overheating and failure will occur for near-complete flow blockage. The radioactivity which then would escape the failed fuel rods would be detected in the steam piping, which is expected to result in a SCRAM twenty seconds after the blockage occurs. GE then *assumes* that the transient will be terminated upon a SCRAM. But there remains the question of whether the overheated fuel will crumble into an uncoolable pile of debris and then heat up due to the afterheat and melt, to trigger steam explosions and a cascade of fuel crumbling. The GE analysis does not treat this obvious question. It considers the worse case of a failure to SCRAM and still predicts no cascading core meltdown.[68] Complete bundle meltdown is predicted, as the bundle becomes starved of coolant; but the GE analysis then argues that the fuel melting will occur in such a fashion (slow drippage, along with the formation of small, separated fragments) that no steam explosions will occur when the hot fuel eventually contacts water coolant. The molten fuel is assumed to eventually melt a hole through the metal duct around the bundle, which is plugged at one end by the assumed blockage, to allow water to leak into the bundle and permanently cool the debris until the reactor is eventually shut down. The debris is assumed to be porous enough to be coolable.[69]

The GE analysis is only as good as its *assumptions*, and several crucial assumptions, though plausible, have not been verified experimentally. Moreover, it is just as plausible to assume that upon melting and crumbling the fuel will pile up into an uncoolable mass of fuel debris, such that a substantial mass of molten fuel could then form—say, 200 pounds—and suddenly contact water to cause a steam explosion, triggering a cascade of core crumbling and melting. As for GE's prediction of no steam explosion, which depends in part on their assumption as to the manner of fuel melting and crumbling, there are no experiments with molten oxide fuel contacting water to support such a hypothesis. According to a report of the Battelle research laboratory, which was prepared for the AEC's Reactor Safety Study (Rasmussen Report), "no completely definitive experiments could be found in the literature." [70]

The AEC's reactor test laboratory, the Aerojet Nuclear Co., has also reviewed the GE analysis and concludes: "Some of the assumptions involved in these calculations are open to question, however, and data is required for verification of the appropriate models." [71] (Aerojet should, however, provide the detailed analysis, including calculations, that supports its questioning of the GE assumptions, and should specify those questionable GE assumptions, so that the public can know the implications, should the assumptions be erroneous.) Thus, all that can be concluded from the GE analysis is that a flow blockage PCMA may be controllable; but this has not been scientifically established, since different plausible assumptions can be made which lead to the opposite conclusion.

The planned PCMA-type testing in the PBF will help to develop pieces of a theory, but because of shortcomings of the PBF, such as the short, three-foot length of the test chamber, several important interactions or processes won't be tested. For example, analysis by PBF scientists "established that with a three-foot test rod it is not possible to attain typical pre-CHF and post-CHF PWR conditions [that is, before and after steam blanketing], simultaneously, in a single test." [72] Therefore, it is doubtful that the existing test facilities can adequately verify DB-PCMA predictions. (Similar shortcomings of the PBF raise questions about the adequacy of planned PBF testing of fuel under PEA and LOCA conditions.) It could be that large-scale reactors will have to be tested. One reason is that, in today's large cores, the fissioning is nearly uniform throughout the core, so that each fuel rod is operated near its limits, which makes cascading fuel rod breakup more likely upon a PCMA. It may be that only large-scale tests can adequately produce spatial near uniformity in the fissioning.

Clearly, what is needed is a thorough study of PCMAs and the experiments necessary to verify the theory. The above-mentioned PBF test plan report only outlines some of the uncertainty of DB-PCMA predictions and is mostly qualitative. What is needed is a full, quantitative evaluation, including theoretical calculations of the implications of the uncertainty, such as the potential for core melting and explosion. Such calculations should be performed for the worse possible PCMAs as well. In this regard, it should be mentioned that the Idaho reactor test laboratory (Aerojet Nuclear Co. and the earlier Idaho Nuclear Corp.) has issued three unpublished, draft versions of a "PBF Test Program Plan." [73] One of the preliminary drafts (IN-1434) includes much scientific "justification" for the proposed PBF testing, which has been left out of the final draft.[74] (Incidentally, the IN-1434 draft treats the PEA- and LOCA-type testing as well.) However, though the drafts provide more discussion of PCMAs (and PEAs and LOCAs), they still fall short of the informational needs outlined above. The NRC should make both preliminary drafts and the final draft available to the public and universities—only the final draft has been mentioned in the open literature.[75]

Design Basis, Loss-of-Coolant Accidents

As for the industry's theoretical predictions of successful ECCS performance in the event of DB-LOCAs, these too have not been verified experimentally by reactor loss-of-coolant tests, though the

NRC is about to conduct one, but small-scale and limited to a PWR, known as the loss-of-fluids test (LOFT).[76] As mentioned earlier, the ECCS appears to be reasonably well designed for the highly specific DB-LOCA (see pp. 23–24); however, there remain numerous, detailed questions,[77] which can be resolved only by this type of testing. For example, will a significant number of fuel rods shatter under the cold shock of emergency coolant? After the LOCA blowdown, the fuel rod cladding is predicted to heat up severely ("red hot") due to the afterheat and the "stored heat" in the interior of the fuel rod. The quenching by the emergency coolant might cause the hottest rods to shatter into an uncoolable heap if they remain too hot for too long a time before being quenched. It is predicted that the AEC's ECCS design criteria will preclude this possibility; but a question evidently remains, due to an alleged experimental inadequacy. This author is not presently in a position to evaluate this concern; but it should be noted.

Probably, LOFT will settle many of the outstanding questions. But there are some which it cannot settle. For example, will the emergency coolant injected by the ECCS reach and cool all of the fuel in the core; or will steam formation and complicated coolant flow patterns form to block or deprive a region of the core of adequate cooling? The use of a coolant channel duct enclosing each fuel rod bundle in a BWR and other BWR features promote predictable flow patterns by channeling emergency coolant through each fuel rod bundle. But in the PWR the situation is less certain because of the lack of such coolant ducts, which would permit lateral coolant flow patterns and possible diversion of coolant away from a hot region of the core. The actual flow pattern, therefore, would seem to depend heavily on the three-dimensional and full-size features of a PWR. The true flow patterns, however, cannot be rigorously predicted from principles of fluid dynamics, because of the complex geometry of the reactor core. Consequently, the DB-LOCA theory for PWRs is necessarily built on hypothesis,* which can be tested only by full-scale LOCA/ECCS reactor tests. Yet, the LOFT core is only about one-eightieth the volume size of a commercial PWR core. Therefore, even if the DB-LOCA theory for the PWR were verified by the LOFT tests, or adjusted to force agreement, it could not be reliably extrapolated for use in predicting the ECCS performance of a large PWR. Moreover, the many

* This author has not rigorously reviewed the "theoretical models" used in the DB-LOCA calculations, and therefore he cannot now vouch that there are no essential shortcomings in addition to those noted in this present work.

crucial design parameters for the LOFT tests, such as the initial fuel temperatures, the core fission "power density" prior to a LOCA, and the core height-to-diameter ratio, cannot all be made equal to the corresponding parameters of a large PWR, due to the nature of small-scale reactor tests, which further questions the validity of extrapolating the LOFT results to large PWR conditions.

Another very serious shortcoming of LOFT, hitherto unmentioned, is that it will be performed with a core composed of fresh, unirradiated fuel rods, when the fuel is strongest, and will *not* also test a high burn-up core, when the fuel will be weakest (and when the radioactivity content of a power reactor would be the greatest).[78] Inasmuch as the fuel rods during a LOCA will be exposed to mechanical forces, severe overheating, and strong quenching and associated thermal stresses and strains in the fuel cladding, and since the object of the ECCS is to prevent fuel rod crumbling, to ensure against a core meltdown, the strength of the fuel rods assumes primary importance. Therefore, it is necessary that a loss-of-coolant test include a high burn-up core as well; though this is probably impractical in the existing LOFT facility, due to the intense radiation emanating from an irradiated core, which would require remote assembly of the test. Moreover, it would take several years just to preirradiate a test core. Unless this were done, however, the public would not have any reasonable assurance of the controllability of a DB-LOCA, even if the presently planned LOFTs of the ECCS concept are successful. Plans for LOCA testing of fuel rods in the PBF have shortcomings, as mentioned earlier (p. 61), and may not be adequate.

Furthermore, LOFT evidently will not investigate the possibility for "ballooning" of the fuel rod tubing (cladding) and resulting blockage of coolant passages within the fuel rod bundles, since the fuel rods in LOFT will not be internally pressurized as typical fuel rods evidently will be.[79] (Ballooning occurs when the fuel cladding heats up in a LOCA, which makes the cladding plastic. Any internal pressure then can blow up the cladding like a balloon, much like blowing glass bottles—as has been found in tests—since the coolant pressure outside of the fuel rods will drop with the loss of coolant.) Fuel rod ballooning, if it occurs, is predicted by the nuclear industry to occur randomly in the core and with limited clad bulging, so that it will not cause serious blockage of emergency coolant to portions of the fuel. While this view is persuasive, the issue may not be settled, except by full-scale tests using commercial fuel, since normal production and use might generate fuel conditions which could lead to coherent rod ballooning in the event of a DB-LOCA, by some unforeseen mechanism.

Moreover, no test of the LOFT type is planned for the BWR. Since the BWR core operates at about one half of the "power density" of a PWR core, (core heat output ÷ core volume)[80] it might be assumed that a PWR test would be more severe and that LOFT is, therefore, adequate. But there is a *reactivity question* concerning the BWR DB-LOCA, due to the large amount of steam bubbles at full power in the most reactive region of the BWR core (above the control rods). In the event of a DB-LOCA, especially a steam line break, the sudden depressurization of the coolant might conceivably lead to a momentary frothing or swell of the coolant in the core in such a way as to move more coolant (water) into the reactive region and thus raise the reactivity and trigger a severe power excursion, even before the SCRAM could take effect. (Recall that the DB-LOCA assumes that the SCRAM occurs.) Such an autocatalytic power excursion is somewhat doubtful. However, present BWR-LOCA theory includes certain approximating mathematical assumptions which preclude a meaningful theoretical answer to this question.[81] Regardless of any theoretical analysis, experimental verification would still be necessary if BWR-LOCA calculations are to be scientifically established. And since full-scale reactors are necessary when investigating reactivity effects, a BWR LOFT would have to duplicate a full-size commercial BWR.

Other thermal-mechanical-fluid considerations tend to call for full-scale DB-LOCA tests in any event. In the recent American Physical Society (APS) review of the safety of water-cooled reactors, it was concluded that even if the DB-LOCA theory was verified by LOFT (for the PWR) there are various factors of the theory which cannot be reliably *scaled up* to apply to full-size reactors and that, therefore, a large-scale LOCA test would be needed, if alternate safety measures to cope with a DB-LOCA, such as underground placement of reactors, are not practical.[82] It seems clear, therefore, that full-scale DB-LOCA tests would have to be performed—say, about sixteen full-scale tests, to cover various combinations of conditions (two reactor types, three pipe rupture sizes, two or three fuel burn-up levels, and two pipe rupture locations, inlet and outlet). About ten years would be needed to build two test reactors (a PWR and a BWR), preirradiate the test cores, and conduct the tests, which would cost an estimated $3 billion.

Another question as to the controllability of the DB-LOCA concerns the integrity of the containment as it is pressurized by the steam of the LOCA. Though it is designed to maintain its integrity, this too has never been fully tested.

Conservatism in the DBA Calculations

Now, it is true that the design basis accident calculations include some *conservatism.* That is, some of the numerical values for the various input quantities are chosen so that the calculational result, mainly, the peak fuel temperature, will hopefully err on the safe side. The AEC has relied on this conservatism when authorizing reactors. To estimate the magnitude of this conservatism is difficult, since it is so diffused throughout the calculations; though the industry has calculated it for the DB-LOCA.[83] This author's rough estimates of the conservatism are given in *Table 3* and average about 40% (the NRC should generate more precise estimates). It must be added that the conservatism does not represent any *positive* safety margin, but is only a way of accounting, subjectively, for the uncertainty in the various quantities to which the conservatism was affixed and for the overall uncertainty arising out of the lack of full-scale experiments. For instance, the NRTS has noted that the theoretical predictions of the fission heat produced in nondestructive power excursions in small reactors, which have been tested, are accurate to only 20 to 50%;[84] that is, the error can be as high as 50%, which presumably consumes whatever conservatism presently exists for DB-PEAs. This does not account for the additional, and unknown, error in predicting power excursions in large cores (for example, "space-time kinetics" effects). Therefore, though the conservatism may provide some assurance that the DBAs will be controllable, it does not *scientifically establish* controllability, as only an experiment can do that.

It is the lay public who must ultimately assess the safety of reactors; and while the layman can understand full-scale, integrated experiments, he cannot practically assess the adequacy of

Table 3 DBA Conservatism Estimates

Accident (DBA)		Estimate %	Comment
PEA			
	PWR	50	Experimental failures indicate this conservatism is insufficient.
	BWR	20	
LOCA	PWR, BWR	50	No integrated tests even on small scale.
PCMA			
	PWR	30	No experimental data at all.
	BWR	10	

complex conservative factors that are embedded in highly esoteric theory. The conservatism is especially unconvincing for the DB-PEAs in view of the above-discussed experimental failures, which indicate that even without the conservatism the accident calculations couldn't confidently predict that the power excursions would be controllable. Furthermore, there are the before-mentioned shortcomings of the small-scale and isolated-effects experiments, which further question the conservatism. Finally, the conservatism pertains only to the very narrowly defined design basis accidents, not to the *worse possible accidents*, the risks of which society must assess as well.

Worse Possible Accidents: Experimental Impracticality

In full-scale tests of DBAs, if we believe the present safety analyses of the nuclear industry, only the core would be expected to be damaged and replaced after each test, allowing reuse of the rest of the reactor. But in destructive full-scale testing to investigate the explosion and radioactivity release potentials of the worse possible accidents (WPAs), the reactor vessel is likely to be destroyed, which would drive up the cost of each test from, say, $20 million per test (core costs) to, say, $300 million to replace the reactor system and for radioactivity cleanup, if cleaning is feasible. This, plus the fact that there are many substantially different WPAs requiring tests, makes any meaningful experimental program economically impossible. Several maximum conceivable power excursions and LOCA experiments using full-scale underground reactors to determine the upper bound of the accident hazard might be conducted at a cost of, say, $1 billion to $3 billion; but these tests may be too hazardous in themselves.

We can at least perform theoretical calculations of all of the WPAs, and include as many detailed effects and interactions in the calculations as we can, within the limits of computer and mathematical capability and our knowledge of the basic physical and chemical processes. Though unreliable, such calculations would still provide useful insight as to what might be possible in an accident. Small-scale, whole reactor destructive testing could also be performed, which would provide substantial information about such problems as autocatalytic reactivity effects, the steam explosion potential of PEAs, and cascading fuel rod failure in PCMAs. In addition, large-core neutron dynamic experiments short of explosion can be used to investigate the validity of the space-time kinetic theory used to predict PEAs. Such limited experiments would reduce much uncertainty in any theoretical model, though the matter of autocatalytic reactivity effects, for example, would not be completely covered.

Reactor Safety Myths

Several myths about the reactor accident potential can now be disposed of. First, there is the unfounded claim that reactors cannot explode.[85] The AEC's *Water Reactor Safety Program Plan,* which is a fundamental reference for all reviewers of reactor safety, furthers this myth by asserting, without citing any references, that in the SPERT power excursion testing of a reactor fueled with uranium oxide, the type of fuel now in use, the core failed to explode "in the severest test that could be performed." [86] (This was the "SPERT-I oxide core test.") However, the *Safety Program Plan* neglected to mention that the oxide core test was a dud, due to a premature rupture of two faulty, "waterlogged" fuel rods,* which terminated the power excursion by expelling some hot fuel powder and coolant (negative reactivity feedback) *before the fuel could be excursion-heated to the melting point, the condition for steam explosion.*[87] The SPERT fuel rods contained UO_2 powder, unlike solid UO_2 pellets used in power reactors. Powdered fuel can more easily be dispersed into the coolant surrounding the fuel rods, leading to rapid boiling and expulsion of coolant. This difference in fuel further diminishes the value of the SPERT test.

Previous testing of two small reactors using *metallic* fuel had resulted in strong explosions, which destroyed the reactors, when the test excursions produced fuel melting.[88] These were the BORAX-I and SPERT-ID tests (see fig. 10 for the BORAX explosion).[89]

The force of the BORAX explosion was equivalent to about 2 pounds of TNT.[90] (A piece of equipment weighing one ton was thrown thirty feet into the air.) However, if we scale this up to present, large-size, oxide-fueled, water-cooled reactors, we can conclude that a 1,000-pound TNT-equivalent explosion might be possible. For comparison, this exceeds the estimated TNT level

* *Waterlogging* means that water seeps inside a cold fuel rod through a crack in the cladding prior to a power excursion test. During the excursion, water would explode into steam, because of its intimate contact with the hot UO_2 fuel within the rod, thereby rupturing the rod, as occurred in the test. Incidentally, the amount of water inside a waterlogged fuel rod is minute and thus, by itself, has a negligible effect on the reactivity in an excursion.

Figure 10. A destructive nuclear runaway (power excursion) test conducted in 1954 of a small nuclear reactor—a 1/500 scale boiling water reactor. This test posed no serious hazard to the public, because little radioactivity was involved and the reactor was located in an Idaho desert. The core was small and operated a short time at a low power level prior to the test; so there was little buildup of fission products (no plutonium was used). Also, the fuel was metallic, which minimizes fission product release due to lower melting temperature, and only a fraction of the fuel melted (no afterheat was involved).

that would rupture either the reactor vessel or its containment.[91] (See table 4.)

Also, the "destructiveness" of this BORAX test was "somewhat unexpected." [92] That is, prior power excursion testing with the BORAX-I reactor showed vigorous "water expulsion" with some damage to the fuel; but no explosion or fuel melting occurred.[93] The destructive test was then conducted by triggering a stronger power excursion that was designed to produce fuel melting and thereby damage the core beyond further use. However, only more vigorous water expulsion, not the explosion that resulted, had evidently been expected. This underscores the necessity for integral reactor experiments to verify theory.

The SPERT-I oxide-core test was designed similarly to produce fuel melting, in order to investigate the steam explosion potential of UO_2 fuel, which could be more severe than metallic fuel due to its higher melting temperature. But the above-mentioned faulty fuel prevented a valid test. The test report concluded:

> The results of these tests do not, however, lead to the conclusion that a 2.2 or 1.55 msec period test [the technical description of the tests*] could safely be performed without damage to the core if the defective rods had been removed prior to these tests. The failure of the defective fuel rods resulted in the earlier shutdown and limited the energy releases to smaller values than expected. Consequently, if the fuel rods had not failed from waterlogging, the tests would have resulted in larger energy releases, and different effects might have been observed.[94]

No explanation has ever been given in the technical literature of why the SPERT-I oxide test was not repeated with the faulty fuel

Table 4 Rupture Level (lb. TNT)

Structure	PWR	BWR
Reactor vessel	95	160
Containment	900	100

* The reactivity was raised initially to about 2% in the severest test, which was the maximum reactivity that the SPERT reactor facility could produce. This corresponds to the power level rising exponentially in the initial phase of the excursion by a mathematically significant 172% in every 1.55 msec time period, which explains the technical description of the test.

problem corrected. Also, the core of the test contained a mechanical "grid" which restrained the fuel rods from bowing inward in order to suppress a strong autocatalytic reactivity effect that was observed in previous testing of the core.[95] If the object of the test was to generate the strongest power excursion that could be performed, then why wasn't the grid removed? Presumably, a worse excursion than that tested could have been generated as a result.

Clearly, the unqualified reference in the *Safety Program Plan* to the "severest test" of the SPERT-I oxide core implies that the presently used oxide fuel is inherently safe against power excursions. Yet, the rest of the PEA section of the plan contains many assertions of theoretical and experimental shortcomings and recommendations for further studies, the thrust of which is that the maximum power excursion potential of water-cooled reactors has *not* been established. In failing to qualify the SPERT-I test results, however, the plan makes the fact of the unestablished power excursion potential seem relatively unimportant. Moreover, the SPERT-I reactor was small (1/250 the size of today's large reactors) and unpressurized; it therefore could not have investigated large-core effects, namely, space-time neutron dynamics, and those autocatalytic effects or their degree which depend on the size of the core.

There was another small-scale, oxide-fueled test reactor, SPERT-III, which could also have been destruct-tested, by causing fuel melting in a power excursion; but it wasn't. Only relatively mild excursions were tested; fuel melting or any other type of fuel damage was deliberately avoided.[96] SPERT-III, shown in figure 11, was a pressurized reactor using high-temperature, high-pressure coolant and was capable of producing steady power under BWR or PWR conditions—that is, with or without steam bubbles in the core—prior to a test power excursion.[97] (In SPERT-I, on the other hand, the coolant was unpressurized and cold.) Destructive excursion testing of SPERT-III, therefore, would have yielded much information on the steam explosion potential of high-temperature coolant.

More importantly, the SPERT-III test reactor could have been used to investigate to some degree the possibility of autocatalytic reactivity effects occurring during moderately severe power excursions—effects which could make a power excursion much worse. An example would be the rapid compression of steam bubbles throughout most of the core, which could be caused by a reactor pressure surge created by an explosion of a limited portion of the reactor core. (This process was discussed earlier, pp. 28–30.) Indeed, such testing was actually planned for the SPERT-III re-

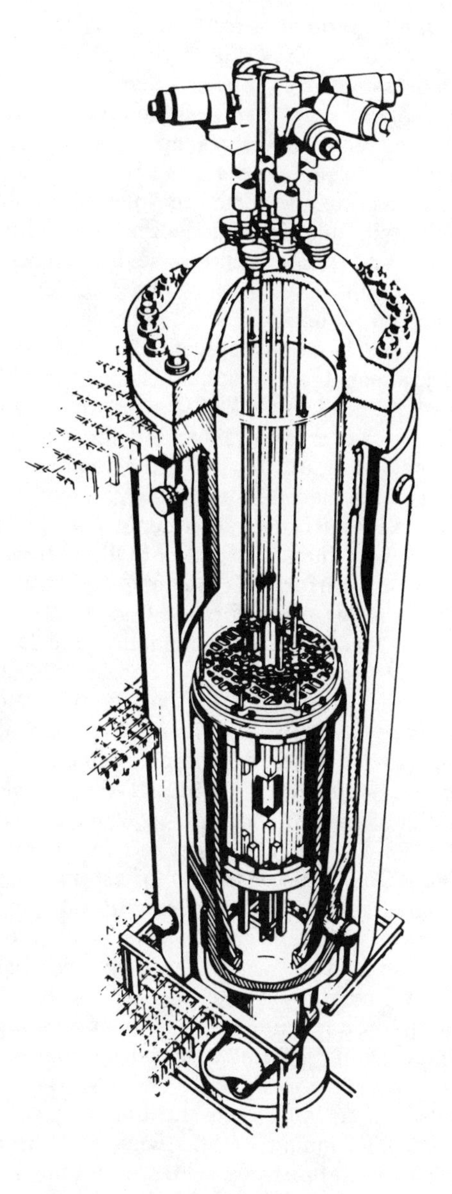

Figure 11. SPERT-III Reactor Vessel Assembly

actor, under the heading of "integral-core limited-damage experiments," according to an "internal report" of the NRTS, dated 1966.[98] However, the experiments were never performed.

Autocatalytic reactivity rises due to coolant changes, such as steam-bubble compression or collapse, are much more dangerous in a uranium-oxide fueled core than in a metallic fuel reactor like BORAX-I, because metallic fuel gives up its fission heat to the coolant very rapidly during a power excursion. The resultant rapid water coolant heating and formation of steam bubbles, during a power excursion, quickly reduce the reactivity to "shut down" the excursion and thereby limit its severity. This shutdown process acted to limit the explosion of the BORAX test. Oxide fuel, however, gives up its fission heat slowly, so that the coolant heating shutdown effect is much delayed. Consequently, an autocatalytic reactivity effect due to coolant changes, such as steam bubble collapse, would not be checked (reversed) by coolant heating (boiling) during a power excursion; and the excursion could then continue virtually unmitigated until the reactor exploded.

Therefore, if the planned "limited-damage" experiments had been performed in SPERT-III, the result might possibly have been an explosion more severe than that of the BORAX test—this may be why the test was not performed. (Remember that autocatalysis may be more serious in large reactors, so that small-core SPERT-III tests would not be sufficient.) Incidentally, the fact that the SPERT-III "limited-damage" experiments were actually planned is not revealed in the open literature—not even in the *Water Reactor Safety Program Plan.* The NRC should make public all internal documents related to the SPERT program, especially those describing the plans and capabilities of SPERT-III relative to investigating the potential of reactor accidents in water-cooled reactors.

That the new Power Burst Facility can investigate PEAs is another myth. The AEC's director of reactor safety research characterized the planned testing in the Power Burst Facility (PBF) as "integral experiments,"[99] implying that the calculational theory for power excursions will be given an integrated test with all known processes and conditions present; but the PBF is a small reactor designed to produce limited power excursions, without core damage, while testing several fuel rod specimens placed in the core for their failure thresholds. (The core is designed to increase the neutron density around the fuel test specimens, in order to increase the fissioning within them, and the specimens will have higher concentrations of fissionable uranium, which together ensure their overheating.)[100] Obviously, the PBF cannot investigate space-time kinetic effects of large-core PEAs or reactivity effects of fuel rod failure in a PEA, since only a few test rods (about twenty-five) would be tested versus the more than 30,000 fuel rods making up a large reactor core. Nor does it seem that the PBF can produce the proper level of afterheat. Indeed, even the originally planned excursion testing of fuel rod specimens in the PBF has been curtailed, in favor of tests related to the DB-LOCAs and PCMAs.[101] Likewise, the highly vaunted Loss-of-Fluids Test Facility (LOFT) cannot be used for large-core PEA investigation, due chiefly to its small-size core; nor, of course, is it planned for such use.

Five

Accident Probability and Informational Needs

THOUGH THE multiple-failure WPAs are not analyzed in the SARs, extensive measures are taken to attempt to make their probability of occurrence negligible, such as back-up equipment, interlocks, and so on. These preventive measures are praiseworthy and are perhaps the best that can be done. Adding on still more safety equipment could possibly reduce the present level of safety (increase the accident hazard), as the plant would be made more complex to operate and maintain. Unfortunately, however, low estimates of accident probability and vague judgments of negligible WPA probability, such as those presented in the AEC's Rasmussen Report, cannot be proven, since they are essentially personal, subjective judgments. The probability of a WPA may not be high, but is it low enough, commensurate with the potential for disaster? (An accident occurring once every hundred years would surely be intolerable if the consequences could be of the magnitude estimated in chapter 1.) Any reactor accident probability figures are based on guesses of the component probabilities and other subjective assumptions, despite the complicated mathematical exercises and data on failure rates that accompany the probability analyses. There is virtually an infinite number of ways in which things can go wrong, due to the extreme complexity of nuclear reactor systems; and there is the unquantifiable factor of human error and carelessness.

As mentioned before, multiple failures are the usual way accidents occur. To illustrate, the BWR manufacturer, General Electric Co., testified in an AEC reactor licensing hearing in 1971 that the probability of the SCRAM system failing is one out of 10 billion.[1] Yet, the SCRAM system of a GE BWR was found "totally inoperative" in a routine check, due to a common multiple failure of a set of electrical devices, which were originally chosen on the basis of the best record of reliability.[2] Furthermore, the AEC revealed to the public that the SCRAM system was "inoperable" on a "number" of occasions at various reactors.[3] Recall that many WPAs would occur upon a SCRAM failure; yet, today's reactors have no back-up SCRAM systems. In September 1973 the AEC issued a policy change requiring a back-up SCRAM system, but only for reactors whose construction permit applications are made after October 1, 1976, which would leave over 200 reactors with no back-up SCRAM system.[4] Moreover, the required SCRAM system will not be a true back-up system, since the design of the control rods and their drive mechanisms will remain the same as in the primary system.[5] (The existing control rods will simply be divided into two groups—one for each SCRAM system.) The designs should be different, however, to minimize the possibility of both systems failing due to a common fault, which is a real possibility, as underscored by the BWR SCRAM system failure mentioned above and by the occurrence of defective control rod drive mechanisms (see p. 38). The EBR-I and Hanford reactors were protected in serious mishaps by the action of a true back-up SCRAM system (see app. 2), but to provide this for PWRs and BWRs would require a major redesign of the reactors.

Indeed, one wonders whether the AEC's recent *Reactor Safety Study* (Rasmussen Report) will be used to rescind the back-up SCRAM requirement, as it claims that the risk of serious accidents in present reactors without a back-up system is already much lower than other risks accepted by society.

As another illustration of the ways things can go wrong, the Vermont Yankee BWR went critical accidentally during a refueling[6] and almost suffered a PEA *with the reactor vessel head off and the containment open* (the reactivity rose close to the delayed neutron fraction, the power excursion threshold). Fortunately, a SCRAM occurred automatically;[7] and even if it hadn't, the excursion would probably have been not severe. However, the PEA no-SCRAM head-off situation has never been analyzed for the course it could take. Moreover, in view of the large excess reactivity, the Vermont Yankee incident could have occurred under worse circumstances and caused a disaster for New England, since any release of radioactivity from the core would have had a direct access to the atmosphere, except for a building roof, which presumably could have been easily breached by an explosion (see app. 2, no. 7). The incident occurred because of violations of procedures, compounded by a breakdown in communications between two work shifts. (The AEC proposed to levy a $15,000 fine on the utility.) [8]

There have been other near-accident incidents, such as: the steam generator falling at Shippingport, which could have caused a LOCA if it had happened when the reactor was operating; the Fermi meltdown; a reactivity incident at Millstone Point, saved by a SCRAM; the totally inoperative SCRAM system in a BWR; and the Dresden partial loss-of-coolant incident.[9] (A description of fourteen such accidents and near accidents is given in app. 2.)

In regard to the design basis accidents, there is no clear probabilistic dividing line between them and WPAs, as shown in the following illustrations.

1. The DB-PEA for the PWR assumes that a single CRDM housing ruptures, which causes a control rod ejection accident; but if one were to fail, it would seem reasonable to assume that adjacent CRDM housings were on the verge of failure as well, due to a common defect. In that case, couldn't the explosive effects of the original failure cause additional CRDM failures in a cascade manner, producing a severe PEA? This possibility has not been addressed in the industry's safety analysis reports.

2. The DB-LOCA is defined as a coolant pipe rupturing spontaneously, due to some faulty condition of the piping, while the reactor is operating steadily at full power. However, a LOCA could just as likely occur when the reactor experiences a pressure surge due to an overheating transient, such as a power excursion or a loss of heat removal from the reactor coolant while the reactor is at full power, which could cause the coolant pressure to reach as high as 3,000 to 7,000 psi in a PWR, compared to the design pressure of about 2,500 psi.[10] (The faulty piping could be on the verge of rupture at normal operating pressure and then break when the first pressure surge occurs.) The ECCS is not designed for LOCAs in which the core is hotter; indeed, the core could be too hot for the ECCS to control.

3. In a DB-LOCA, the steam generator tubes are assumed to remain intact after they become empty of reactor coolant when it blows out of the reactor system. But if only 5 out of about 14,000 steam generator tubes should break,[11] high-pressure steam formed from the hot, pressurized boiler water in the steam generator would flow into the empty reactor coolant system through the broken tubes and eventually into the core by way of the coolant piping, blocking the emergency coolant from reaching and

cooling the core (refer to fig. 1). This serious possibility could cause a potentially disastrous core meltdown [12] and is underscored by the frequent steam generator tubing malfunctions (cracking-and-thinning deterioration and leakage) that have occurred in PWRs.[13] Though the tubing is designed with a three- to sixfold safety factor against collapse and is designed to withstand vibration stresses during a LOCA, these assurances assume no deteriorated tubing.[14] Therefore, it would seem prudent to assume that at least ten of the tubes break in a LOCA. (Incidentally, the public has yet to hear the industry's response to this concern.) [15]

4. In the design basis, control rod dropout accident (BWR), the reactivity "worth" of the dropped control rod, that is, the reactivity increase potential, was at one time assumed to be 2.5%, which was calculated to produce a power excursion with a peak energy yield less than 280 cal/gm, the AEC's safety limit.[16] Subsequent and evidently more refined calculations showed that the safety limit would be greatly exceeded,[17] and as a result, the control-rod worth assumption for this DB-PEA was reduced to 1.4%, without any apparent change to the reactivity worth potentials of the control rods.[18]

5. As a final example, the design basis flow blockage accident for a BWR fuel rod bundle assumes that only *one* fuel rod bundle is blocked. Yet, it would seem prudent to assume that at least two bundles are blocked, since this has already happened in another reactor,[19] where fuel melting might have led to a PEA if a SCRAM had not occurred.[20] (However, this reactor was of an entirely different design, so that the incident sheds no light for the BWR.)

From the foregoing examples, it appears that the DBAs are selected somewhat arbitrarily, and at times purely on the basis of whether unavoidable, state-of-the-art calculations can show no violation of decreed safety limits rather than on well-justified grounds. We must remember, too, that the single-failure DBAs have not been scientifically established to be controllable, which adds to the uncertainty of the probability of disastrous accidents.

The question naturally arises: If there are so many ways serious accidents can occur, then why haven't any occurred? It must be answered that, at least over the past seven or so years, the limited number of reactors have been operated basically as designed in regards to safety. Whether this record can be kept up indefinitely is a matter of personal judgment. Perhaps nuclear reactors can be and are being made and operated safely under established regulatory requirements. This is what we must decide, and the only way to reach a sound decision is to study the accident hazard in some detail; to learn the various ways accidents can occur, the potential for explosion and public harm, and the history of actual malfunctions that have occurred (for enough malfunctions occurring together can produce certain kinds of accidents); and then to reflect on general human experience and make a judgment.

In regard to malfunction occurrences, there have been near-accident situations and incidents which we should study to help form a judgment (see app. 2 for a discussion of some of these), but a full evaluation of the history of actual reactor malfunctions has yet to appear. The Rasmussen Report, for example, draws only on a selected period of time for malfunction occurrences.[21] Indeed, instances of human error which have actually occurred in reactor operations and mishaps are not analyzed.[22] There is a great deal of information on malfunctions already compiled in other technical reports, examples of which are given throughout this book; but the information is not complete, for the reports make the qualification that only "selected" occurrences or "representative samples" are reported.[23]

It is therefore suggested that the NRC should investigate all nuclear power plants for instances of reactor malfunction and maloperation, tabulate all occurrences significant to the possibility of serious accident (sorting them out according to specific accident possibilities), and show their significance. Such full, objective information is needed in order to rationally assess the matter of accident likelihood.

The fact that no disastrous accidents affecting the public have occurred in the United States shows the obvious care given to reactors in design and operation, and not that reactors can survive successfully a DBA or a WPA situation, since no such situations, fortunately, have ever occurred. This safety record provides very little assurance that the probability of a disastrous accident is acceptably low, however, because there have been only about 200 reactor-years of operating experience (roughly thirty reactors operating for an average of six years), which compares poorly with 40,000 reactor-years that would be consumed by the 1,000 projected plants in just one forty-year period—the design lifetime for reactors. (Actually, only about 130 reactor-years have been accumulated by the large reactor plants—400 MW or larger—and this averages out at about three years of operation per reactor for the forty-five large plants presently in operation.)[24] Furthermore, it is an engineering fact that reactors will become degraded with use, due to metal fatigue, corrosion, embrittlement, and so on, so the three years of average reactor use accumulated so far by the large reactors provides even less assurance. This is underscored by the cracks found in the large, high-pressure ECCS pipes (at the vessel connection), which prompted the NRC to shut down twenty-three BWRs for inspections,[25] and the cracks found in BWR control rod drive mechanisms (see pp. 38–39).

It is contended, therefore, that the only way that society will be able to make a sound decision as to whether nuclear reactors are safe is to act on the basis of full, *objective* knowledge of the accident hazards and associated scientific information, not on the basis of a partial analysis and vague, *subjective* assurances of the adequacy of theory and the remote probability of serious accidents, given by those who have vested interests in sustaining nuclear power. The necessary information would take the form of answers to the following questions:

1. What are all of the accident possibilities and possible consequences? *All* reactor accident possibilities must be analyzed and calculated (as best we can) for the course that each could take, including all conceivable autocatalytic processes and other adverse factors that cannot be ruled out.[26]
2. What are the specific theoretical uncertainties?
3. What theories have not been verified experimentally?
4. What chain of failures would have to occur to cause each accident? All experience of reactor equipment failures and human error related to each accident chain should be summarized, so that the public can best judge the likelihood or remoteness of each accident.

With this information, the decision makers for society will be able to make a sound judgment, for they will know the full extent of the risks, including the uncertainties. They might select as a minimum those particular reactor accidents which must be controllable; or they might reject nuclear power altogether on the ground that the risk of worse possible accidents is unacceptable or that needed experimental verification of accident calculations is impractical. If society wishes to consider nuclear power further and to select those accidents which must be controllable, then full-scale reactor experiments would have to be performed to verify the associated accident predictions, if this is practical. If experiments are impractical, society will then be confronted with the choice of rejecting nuclear power or taking an ill-defined chance that the selected accidents will be controllable in theory but not in established fact.

Six
The Rasmussen Report

IN AUGUST 1974 the AEC issued a report of a reactor safety study which purports to consider all accidents but in fact does not. This is the so-called Rasmussen Report, named after the MIT nuclear engineering professor, Norman C. Rasmussen, who chaired the $4 million, fifty-man, three-year *Reactor Safety Study*.*[,1] The report makes an important contribution by analyzing some reactor accidents that are worse than the DBAs. Specifically, the report treats: (1) the LOCA-without-ECCS type of core meltdown; (2) a class of accidents called "transients," which basically come under this author's category of power-cooling mismatch accidents (PCMAs) and heat exchange accidents; and (3) the spontaneous reactor vessel rupture. The power excursion accidents (PEAs) and the worst forms of PCMAs are essentially excluded. Moreover, on the basis of certain assumptions, the report treats the "transients" and the reactor vessel rupture as no worse than a LOCA without ECCS (slow core meltdown), which is not the worst course these accidents could take. (The report's treatment of the LOCA-without-SCRAM-type accident was examined earlier; see pp. 34–39.) In short, the report is grossly inadequate in scope. This fact, plus other crucial and unsubstantiated assumptions to be demonstrated below, explains why the estimates of maximum accident consequences given in the report are minor compared to the extrapolations from WASH-740 given at the outset of this synopsis. Specifically, the maximum Rasmussen Report estimates

* The report was published though it was labeled a draft. The final report, dated October 1975, is commented on in app. 1.

are: (1) a lethal range of the radioactive cloud of only 15 miles; (2) evacuation and serious agricultural restrictions for 400 and 4,000 square miles of land, respectively (though it should be noted that the last figure amounts to about one half of the size of Massachusetts); and (3) restrictions on milk for a few months due to radioactive iodine over 40,000 square miles (an area the size of Ohio).[2]

Omission of Power Excursion Accidents

Professor Rasmussen has asserted that the report considers "the worst accident imaginable," [3] and his report states that "it covered all potential accidents important in determining the public risk." [4] These assertions are patently not true, since the report does not treat the severe PEAs. The report merely mentions some PEAs in a table (for example, the BWR control rod ejection and the PWR cold water accidents, described earlier),[5] but does not analyze and calculate them for their probability of occurrence and potential for explosion and radioactivity release. The report neglects these accident possibilities on the basis of vague assertion that they are so "unlikely" that they are not important "contributors of public risk" and, therefore, need not be treated.[6] These severe accidents are then placed under the heading of "unanticipated transients" or "unlikely [accident] initiating events" and dismissed. Yet, the Rasmussen Report presents elaborate probability analyses of many lesser, though some serious, accident possibilities. Since the report concludes that these analyses prove the risks are acceptably low, one would think that the report would have used the same methods to analyze *all* severe PEAs, in order to "prove" that they are so "unlikely" as to not contribute significantly to the reactor accident risks.

In a table of "BWR Transients," for example, the Rasmussen Report lists the "[Control] Rod Ejection Accident" without any further elaboration or analysis, other than to say that "[s]tructural members are incorporated in the BWR design to prevent" it.[7] This accident happens to be one of the most severe nuclear runaways possible, and only two failures need occur to cause it: (1) a CRDM housing must rupture and break off the reactor vessel; and (2) the associated control rod movement blocking device must malfunction, either by improper installation or by being left uninstalled. Who is to say that the probability of this simple two-fault accident chain is negligible? The blocking device is removable to allow for CRDM repair work. Could a work shift neglect to tell the next shift that they did not install one of the devices, as occurred with other devices in the Vermont Yankee accidental criticality incident? Two early PWR steam generators ripped from their support hangers partly because some bolts were not installed.[8] (The DC-10 airplane crash near Paris of March, 1974 was traced to a door brace not installed.)[9] In short, mistakes and malfunctions do happen; so the Rasmussen Report should have revealed the failure chains of all severe PEAs; presented all significant histories of malfunctions that are relevant to them; and calculated their potential consequences. Consider the opinion of the NRTS: "In attempting to determine the importance of any type of [reactor] accident, it is a common and justifiable procedure to consider both the probability of occurrence and the possible consequences of the accident" (PTR-815, p. 1).

Furthermore, the Rasmussen Report does not explain the power excursion phenomena, nor does it even mention the settled terms "power excursion" or "nuclear runaway." In several places the report admits that a "rapid power transient" could occur, but does not discuss the full significance.[10] For example, the report merely notes that if the steam valve closes in a BWR without a SCRAM, the "power level will increase" (due to the *reactivity feedback* of the rise in coolant pressure, which the report does not clearly explain), the reactor vessel could then "fail," and the core would melt.[11] However, the report does not elaborate with any theoretical analysis or supporting references with which one could pursue the possibilities for autocatalytic nuclear runaway and the question of the fission product release potential relative to a slow core meltdown. Nor does the report pursue these matters. As another example, the report mentions the BWR *control rod drop-out without SCRAM* accident and adds merely that this "could perhaps" result in a failure of the reactor vessel and the containment.[12]

Strontium 90 Release Assumption

The Rasmussen Report assumes only a 6% fractional release of Strontium 90 fission product radioactivity in the worst core meltdown accident treated,* the LOCA without effective ECCS.[13] Sr 90 is one of the most biologically harmful species of radioactivity,

* 10% release from the reactor, with 60% of that escaping the containment. Hence, $.1 \times .6 = .06$, or 6%.

and the main determinant is agricultural land contamination.[14] However, there are no sound fuel meltdown/fission product release experiments to justify the low 6% assumption.[15] A key fuel melting experiment by Browning, et al., of tiny, gram-size cylinders of UO_2 showed up to 99% release of strontium from the UO_2 material, though only 1% escaped the tiny test chamber, due to condensation of strontium vapor on the relatively cold surfaces of the chamber, which tightly surrounded the miniature UO_2 cylinder.[16] The Rasmussen Report and the industry prefer to cite the 1% figure of the Browning experiment instead of the 99% figure, arguing that should any strontium be released from molten UO_2, it will quickly condense on adjacent, presumably less hot, core surfaces, as occurred in the miniature test.[17] The experiment was characterized, however, by a high, maximum surface-to-volume ratio due to the miniature size of the specimen, which maximized the surface condensation of strontium vapor. The ratio in the miniature test was about 600 times greater than in a reactor situation. Therefore, a higher Sr 90 release fraction can be expected in real core meltdown accidents, but the only way to ascertain it is to conduct a full-scale core meltdown test and measure it. Such a test, though on a much smaller scale, was planned (the original purpose of LOFT), but it was canceled.[18] The Rasmussen Report recognizes this deficiency and vaguely includes fuel meltdown tests with a small surface-to-volume ratio in its list of "Research Suggestions" at the end of the report.[19]

Incidentally, the probable purpose of the Browning miniature UO_2 melting test was *not* to fairly simulate a reactor accident but to simply measure the strontium release from UO_2 material upon melting, as distinguished from metallic fuel material. Metallic fuel is expected to release only 1% of strontium upon melting due to its lower melting temperature.[20] Since UO_2 melts at a much higher temperature, a larger release fraction from molten UO_2 can be expected, and it was in fact observed in the Browning test (up to 99%).

Of course, the Sr 90 release fraction will also depend on the severity of the accident. In this regard, the Rasmussen Report's 6% assumption does not apply to severe PEAs.[21] In the absence of definitive experiments, we ought to assume a 50% release fraction—as did the WASH-740 report,[22] though for other reasons—especially since the melting temperature of UO_2 greatly exceeds the boiling temperature of Strontium (5,000° F *v.* 2,100°F), which suggests a full release of strontium from molten UO_2, as supported by the Browning experiment.

Fission Product Dispersal and Contamination Limits for Sr 90

The Rasmussen Report assumes that the radioactivity is released in the form of dust-size particles that are much heavier than the particle size assumed in WASH-740 (125 times heavier or more) [23]; heavier particles tend to fall out close to the plant, instead of contaminating large areas. Also, the Rasmussen Report *assumes* a 26-fold higher Sr 90 contamination limit for agriculture than was *derived* by the WASH-740 report.[24] These less restrictive assumptions appear to explain why the Rasmussen Report estimates of damage consequences are even less than the 1957 WASH-740 estimates. For example, the Rasmussen Report calculates a maximum of 4,000 square miles of land damaged agriculturally due to Sr 90; whereas WASH-740 calculated as much as 150,000 square miles, which is greater than the size of Illinois, Indiana, and Ohio combined.[25] Yet, the total quantity of Sr 90 which the Rasmussen Report assumes is released to the atmosphere is about 60% *greater* than that assumed in the WASH-740 report. (Though the Rasmussen Report assumes a much smaller Sr 90 release fraction, the total quantity released is greater, since today's reactors contain fifteen times more Sr 90.[26])

The *particle size* of the released radioactivity can only be established by full-scale experiments, since it is determined by complex physical and chemical processes; and moreover, it will depend on the severity of the accident—that is, the higher the temperatures and explosion violence, the smaller the particles. As for the 26-fold higher *contamination limit* for Sr 90, the Rasmussen Report offers no justification, except to cite six references,[27] one of which is WASH-740, which is contradictory. Of the remaining five references, one uses a value very close to the WASH-740 value (0.15 microcuries of Sr 90 per square meter *v.* 0.1 microcuries per square meter used in WASH-740), and two do not even address the subject.[28] The fourth reference, an article by Parker and Healy published before WASH-740,[29] does not specify a strontium 90 contamination limit, though it lists a "fallout limit" for total fission products in regard to "crops," from which it is inferred that the Sr 90 limit was implicitly assumed to be 0.5 microcuries per square meter, or five times the WASH-740 value, not twenty-six times. More importantly, the Parker-Healy article does not derive the assumed fallout limit, nor does it cite any supporting reference.

The remaining reference turns out to be the source of the 26-fold higher limit used in the Rasmussen Report.[30] However, this reference is a draft, unpublished, undocumented, and unchecked report written in 1966 that "had no connection whatsoever with a government project." [31] The soil contamination limit is based on some encouraging but undocumented data on low Sr 90 uptake coefficients by one plant (corn), which ignores other foods and food chains. Professor A. Barker of the Plant and Soil Sciences Department of the University of Massachusetts has reviewed the WASH-740 rationale and this Rasmussen Report reference and concludes that the more stringent WASH-740 contamination limit is better reasoned. Indeed, the author of the 1966 report, J. W. Healy, stated that his draft study was "incomplete" and "could be potentially misleading in its present form" and that he was "planning to revise" it,[32] which indicates that the Rasmussen Report is premature in relying on the reference. We may also question whether the Rasmussen Report's higher contamination limit for strontium 90 is based on relaxation of the maximum permissible radiation dose to the general population.

Finally, even if we assume a twentyfold greater contamination limit, a severe accident in which 50% of the Sr 90 is released would still mean that 40,000 square miles would require long-term agricultural restrictions (an area size of Ohio, or Connecticut, Massachusetts, Vermont, and New Hampshire combined), assuming heavy Sr 90 dust particles. If light Sr 90 dust particles are assumed, an estimated 150,000 square mile would be affected. Obviously, the higher contamination limit of the Rasmussen Report does not eliminate the need to know the potential for Sr 90 release in core meltdown accidents—LOCAs as well as PEAs and PCMAs.

Range of Lethal Fission Product Cloud

The Rasmussen Report assumes that of the readily vaporizable, short-lived radioactivity, called the volatile fission products, about 75% is released to the atmosphere in the most severe core meltdown accident treated in the report, the LOCA-without-ECCS-type accident.[33] If we extrapolate from the WASH-740 report to account for the sixfold greater quantity of such fission products in present reactors, we would predict a radiation cloud with a maximum lethal range of sixty miles. (Short-lived radioactivity is much more intense than long-lived radioactivity, since the former decays faster and thereby gives off a faster rate of harmful radiation, which in turn yields a high radiation dose for those persons exposed to a fission product cloud, causing acute radiation sickness and death.*) Yet, the Rasmussen Report estimated only a fifteen-mile lethal range and, thus, a maximum number of "acute deaths" caused by the passage of the cloud of only 2,300. However, the report is vague about whether this is the maximum *possible* consequence calculable by their accident model. The report said only that the worst figures presented are the "peak consequences for very unlikely accidents," which does not definitely mean that the worst possible case is included.[34] It appears to this author that the Rasmussen Commission did not calculate the accident consequence for several adverse meteorological and other aggravating factors that could exist together, as this cannot be ascertained from their report. These would affect the size of land area requiring evacuation or restrictions on living due to radioactive fallout on the ground, as well as the lethal range of the cloud.

Furthermore, the Rasmussen Report estimates that only about 22% of the *total* short-lived radioactivity would be released in the worst accident treated. This is because most of the short-lived radioactivity is not readily vaporizable, as are the volatile fission products. But if the more severe power excursion accidents and power-cooling mismatch accidents, or even a core meltdown in a LOCA (without containment spray in a PWR), could release much of the nonreadily vaporizable fission products as well, such as cerium,† as they might in the case of strontium 90 (see pp. 83–84, 183), then there would be the potential for greater lethal range and area of ground contamination (fallout).

Transients

The Rasmussen Report's treatment of the class of accidents called transients and of the spontaneous reactor vessel rupture is vague and excludes the worst course the accidents in each class could take. Some of the transients treated, which basically would come under the headings of PCMAs and heat exchange accidents, could (1) lead to fuel melting in an intact, fully pressurized reactor coolant system (RCS) [35], which in turn could cause a steam explosion and thereby rupture the reactor vessel; or (2) simply overpressurize the RCS, which also could cause the reactor vessel to rupture.

However, the report neglects the possibility that the reactor

* Or radiation-induced cancer later on.

† In the Browning experiments, the release fraction of cerium was essentially the same as strontium, namely, about 50% to 90%.

vessel will explode (rupture) in such a way as to breach the containment *promptly* by flying vessel fragments (missiles) and thereby allow the whole core to melt down, due to the loss of coolant, in a breached (open) containment. This would allow the fission products to escape directly into the atmosphere through the containment rupture and maximize the release of the fission products. The report expressly admits that such *prompt containment failure* is a possibility but neglects to estimate the harmful consequences. Instead, it assumes that the reactor vessel (or a coolant pipe) will rupture in such a limited way as to cause a LOCA without effective ECCS, that is, a less violent loss-of-coolant situation.[36] (Incidentally, the report is vague about how the ECCS would be made ineffective. If the NRC would elaborate, one could better appreciate the design limitations of the ECCS.) Under this assumption, the containment would rupture after a delay time of about one hour after core melting first occurs, due to one of the several mechanisms for delayed containment failure associated with a LOCA without ECCS, such as steam explosion, hydrogen overpressure, and so on (see pp. 22–23). This would mean less fission product release to the atmosphere.

The reason the Rasmussen Report predicts a high, 70% release of the readily volatile fission products, such as iodine, for the slow, LOCA-without-ECCS type of core meltdown (delayed containment failure) in a PWR is that the containment sprays are assumed to be inoperative throughout the accident as well. This eliminates the washout of fission products and presumably allows the heat of the molten fuel debris to revaporize much of the more volatile fission products that may have condensed out inside the containment when they first boiled out of the core melt. However, less volatile fission products, such as strontium and cerium, would be more difficult to revaporize, once they have condensed out on noncore surfaces within the containment. (In a prompt containment rupture situation, the additional pessimistic assumption of inoperative containment sprays in LOCA-type meltdowns in PWRs is unnecessary.) For BWRs there are no containment sprays, which explains why the report predicts a relatively high (60%) release of iodine for the delayed containment failure case. Thus, we see that prompt containment rupture would seem to maximize the release of fission products by minimizing the condensation and fallout of fission products in the containment.

The report summarizes an elaborate analysis of the LOCA without ECCS to arrive at their worst-estimate category of fission product release (6% of Sr 90 and about 20% of the rest).[37] In contrast, the worst transients treated in the report and the spontaneous vessel rupture are simply assigned to this fission product release category on the basis of a short, vague discussion in which delayed containment failure is implicitly assumed.[38] Even if the containment failure were delayed, however, these accidents would presumably generate higher core temperatures and cause more of the core to melt than the LOCA without ECCS, which would mean greater fission product release. Also, we must remember that the strontium and cerium release assumptions for core meltdown have not been established as an upper limit for the LOCA case.

Another consideration is that a prompt containment failure would allow the hot coolant to be discharged (as steam) to the atmosphere *ahead* of the whole-core meltdown and fission product release (boil-off) processes. Conceivably, this could mean that the fission products would hug the ground as they were carried downwind, instead of being carried aloft by the heat of the steam, causing greater harm to the exposed population. (This conceivable effect would apply to the PEAs as well, which would have a greater potential, plus the likelihood, of causing prompt containment failure.)

The Rasmussen Report does not even describe the transients, except for a vague listing of them,[39] nor are any analyses of these accidents referenced. (This author has identified at least one specific PWR transient that would lead to fuel melting in an intact reactor coolant system, namely, the steady, slow withdrawal of control rods without SCRAM in a PWR. This is a PCMA, rather than a PEA, since the reactivity rises slowly. In general, the PWR transients are PCMAs; whereas many of the BWR "transients" in the Rasmussen Report are PEAs which are potentially severe. The report's treatment of these BWR transients will be critically examined later.) A detailed analysis of the transients would have required that the question of the *prompt containment failure* be dealt with. Instead, the report avoids this by giving only two brief examples of transients which do not involve core melting in an intact, pressurized reactor coolant system,[40] as if these examples are the most severe transients possible, which may not be true. These examples were selected on the basis of dominant" or highest risk. But since the report's definition of "risk"—the accident probability multiplied by the predicted consequences—involves subjective probability estimates, the "dominant risk" transients would draw attention away from those with worse consequences.

The possibility of prompt containment failure by vessel rupture, either spontaneously or in a transient, is noted in several places in the voluminous report, but the question is not pursued. Specifically, the report states:

> However, because of the physical plant layout [PWR], there is some small probability that a large vessel missile could in fact impact directly on the containment and penetrate through the wall. This type of rupture could involve a core meltdown in a non-intact containment.[41]

For the BWR, the report states that "the containment is small in size and its proximity to the reactor vessel would indicate that severe vessel ruptures might either tear or cause missiles to penetrate the containment shell." [42] The report simply leaves it at that, however, without pursuing these possibilities in the included accident charts.[43] Since the reactor vessel rupture is included among the accidents considered, the layman might assume that the report treats the worst possible reactor accident, which is not the case. Again, the report notes that "fuel melting in an intact reactor coolant system (RCS)" could rupture the RCS,[44] but the subsequent discussion is vague and does not explore prompt containment failure.

It should be noted that the Rasmussen Report does identify a great many different transients, mostly heat exchange accidents, that are predicted to be just as severe as a LOCA without ECCS in terms of fission product release,[45] as they produce core meltdown and eventually a ruptured containment. The report assumes that the fission product release would be comparable to a LOCA without ECCS, as noted earlier—70% iodine and 6% strontium 90 radioactivity release. These are the worst release fractions assumed in the report. But even if we assume that each of these transient accidents would take the course described in the report, the *consequences* of these releases could be much more severe than the report estimates, as discussed earlier. Furthermore, the strontium release fraction could be greater (see pp. 82–83). It is of interest to note that the report estimates that the worst transients treated have a combined probability a thousand times higher than a LOCA-without-ECCS accident.[46]

Finally, it should be noted that the Rasmussen Report excludes other transients that are regarded as "unlikely." These too should be treated, including the ways in which they can be caused. Though most of these are PEAs, some are PCMAs. For example, one of the unlikely transients occurs when all of the coolant pumps mechanically seize (stop) at once, which is a severe PCMA. Presumably, this could be caused by an earthquake.

It is concluded that the subjects of transients and reactor vessel rupture are not adequately treated in the Rasmussen Report. The discussion is vague and not specific.

BWR transients, examined more closely. Those BWR transients treated in the Rasmussen Report that are labeled "likely initiating events" are especially important to examine because many of them become PEAs if the reactivity is not reduced by a SCRAM, or by an automatic stoppage of the coolant recirculation pumps, which reduces the reactivity by causing more coolant boiling in the core (see pp. 25–26). These accidental transients involve either a steam valve closure, an increase in reactor coolant flow, or the injection of cold water into the reactor. Such transients either compress, sweep out, or condense the steam bubbles in the core to increase the reactivity. There are many such transients (see table I-14 of the report).

According to the report, which this author has verified for some cases, each of these transients with a failure to reduce the reactivity involves or eventually experiences a steam valve closure at or near full power for various reasons. This causes the reactor pressure to rise, since the steam discharging from the core would then be confined to the reactor vessel. This in turn collapses (compresses) the core steam bubbles, thereby increasing the reactivity, which causes the power to become even greater. This causes the pressure to rise even more, and so on in a spiral. This is the very situation discussed in chapter 3—the BWR *steam valve closure* accident, where this author calculated initial fuel melting occurring within six seconds and a plausible, very severe, autocatalytic nuclear runaway resulting therefrom. (Recall the assumed Roman-candle effect, pp. 28–29 and fig. 9. Incidentally, the NRC staff director of the Rasmussen reactor safety study group, S. Levine, has asserted that this Roman-candle scenario for autocatalysis is very unlikely,[47] but no basis for this assertion has been advanced—that is, no theoretical analysis has been offered.) However, the Rasmussen Report predicts no such nuclear runaway for these BWR transients. Now let us examine the Rasmussen Report's treatment of this unstable pressure-power spiral.

The report assumes that a reactor coolant pipe will rupture in the transient when the "design pressure" of the coolant "piping" is exceeded by the rising coolant pressure.[48] The design pressure of the reactor coolant system (RCS)—that is, of both the piping and the reactor vessel—is 1,250 psi compared to the normal operating pressure of 1,050 psi.[49] This assumed rupture would cause a LOCA, but would at least tend to relieve the coolant pressure and thereby stop the steam bubble collapse/power rise spiral before it could cause fuel melting in an intact RCS, for this author's calculation predicts that fuel melting does not occur until the coolant pressure reaches 1,550 psi. Thus, the less severe course predicted by the Rasmussen Report would require the assumed pipe rupture to occur before this 1,550 psi pressure level is approached.

The piping and reactor vessel are not *designed* to rupture at the 1,250 psi "design pressure," however, as the Rasmussen Report assumes. Rather, there is a safety factor of 2.25 designed into these components, which means that they are not expected to rupture until 2,700 psi. (The purpose of the design pressure, which is about 20% above the normal operating pressure, is to *ensure* that the vessel and piping will *not* rupture during limited pressure surges that are expected to occur during operation.) Indeed, General Electric Co., the BWR designers, has proposed that the 2,700 psi figure be used as a safety design limit for just this kind of "transient without SCRAM."[50] They assumed, however, that at least the recirculation pump turn-off safety feature would work, which would prevent fuel melting by reducing the reactivity and hence the core power level. Still, GE calculated a peak pressure of 1,535 psi for the steam valve closure (without SCRAM) transient, which they argued would not cause any rupture of the piping or vessel.[51] Therefore, the Rasmussen Report avoids the more severe situation of fuel melting occurring in an intact, highly pressurized RCS by simply assuming another course—the arbitrary pipe rupture—contrary to the facts of the probable pressure capabilities of the system.

We can conclude, therefore, that fuel melting will occur in an intact RCS, which might then cause an autocatalytic nuclear runaway. Even if the melting does not cause a nuclear runaway, however, it could, in an intact, extra-pressurized RCS, end in the complete rupture (explosion) of the reactor vessel, instead of a coolant pipe, resulting in a prompt containment failure due to flying vessel fragments. Or, if a pipe should rupture, would a nuclear runaway be triggered anyway, due to an unexpected autocatalytic reactivity effect of the coolant blowdown?

The course of the accident assumed in the Rasmussen Report —the pipe rupture (LOCA) occurring when the piping design pressure (1,250 psi) is exceeded—is, nevertheless, serious and has nuclear runaway possibilities of its own, which the report notes somewhat cryptically, stating that when the ECCS attempts to reflood the core with emergency coolant, the core will return to criticality, because of the SCRAM failure:

> The resulting rapid increase in reactor power, once the core reaches criticality, might result in a reactor vessel failure, or might cause the reactor to "chug" (go from subcritical to some significant power level and back to subcritical) for some period of time. This [chugging] is assumed to eventually result in massive fuel cladding failures and/or a core melt.[52]

Note the references to "rapid increase in reactor power" and "reactor vessel failure," which allude to the possibility that a severe power excursion could be triggered by the reflooding process to explode the reactor and hence cause prompt containment failure. The power level "chugging" would be the only other possibility allowed by the report. To predict just what would be the outcome of this complex reflood/recriticality process would be a formidable, if not an impossible, task, due to the complexities explained earlier (see pp. 16–17, 28–29, 44–58). The report does not reference any analysis or any documentation for these situations. Furthermore, full-scale experiments would be needed to verify any prediction. (There is also the uncertainty as to a power excursion being caused in the blowdown phase of the LOCA; see p. 35.)

Despite these possibilities for severe accidents, the Rasmussen Report assigns the BWR transients-without-reactivity reduction to the fission product release category calculated for the less severe LOCA-without-ECCS accident; for example, 6% strontium 90 release.[53] Clearly, the accidents could cause a greater release.

In regard to the likelihood of these BWR transients, it should be noted that the Rasmussen Report estimates that they "dominate the probability" of all serious accident possibilities in which a heavy radioactivity release could occur.[54] It is interesting to estimate the likelihood of such a BWR transient using the report's probability estimate of one out of five million, that is, .0000002 per reactor per year.[55] Assuming 250 BWRs, this would mean one such accident every 20,000 years. This probability value is based on an assumed 1% chance factor for a catastrophic steam explosion occurring when the core finally melts down after the arbitrarily assumed pipe rupture.[56] But since the transient would more likely follow a more dangerous course (namely, an explosion due to an autocatalytic nuclear runaway or at least excessive pressure buildup), we should assume a 100% chance for catastrophic explosion, which increases the probability to one every 200 years. Then, allowing for a factor of four to account for other uncertainties in the probability estimate, as the report does,[57] we arrive at a probability of one such accident every 50 years, a much greater risk than one every 20,000 years. Moreover, the accident would be more serious. However, neither probability estimate should be regarded as established scientific fact. At the end of this chapter we shall examine more closely the report's probability analysis.

Summary and suggestions. The Rasmussen Report does not justify by theoretical analysis or supporting references that the fission product release fraction (and particle size) for the transients mentioned in the report will be limited to the values assumed, for example, the 6% strontium 90 release value. Presumably, the assertions of what could happen in the event of these severe transients are based on some numerical analysis—that is, some computer simulation of these accidents—but the report cites no documentation. None of these accident possibilities has been analyzed in the industry's SARs or elsewhere in the open literature. If such studies were made public, they could serve as a basis for pursuing the adequacy of power excursion theory, the possibilities for autocatalysis, and the explosion and fission product release potential of these reactor accidents. Further, the method of presentation of accident analyses used in the Rasmussen Report is not practical for review purposes, for it represents a general way of summarizing the many different transient possibilities. Though this is helpful, one really needs a separate booklet or chapter on each individual accident possibility, including all PEAs and PCMAs, which should contain: a description of how the accident develops and the physical processes that could occur, with diagrams and sketches; pertinent data that are obviously important to assessing different possible outcomes, such as power excursions; a discussion of possible or plausible outcomes; summarized theoretical analyses and documented detailed analyses; a discussion of the experiments or lack of such related to the accident; and a tabulation and discussion of the significant abnormal reactor operating experiences (accidents, near accidents, and equipment failures) that are relevant to assessing the probability. With such information, physicists and engineers in and out of the industry would then be able to systematically assess the accident hazard of nuclear reactors, including the potential for autocatalysis and the experimental needs and scientific uncertainties.

Neglect of Cascade Fuel Failure

The Rasmussen Report neglects to treat the class of conceivable PCMAs that would involve a cascade of fuel rod failures occurring when the reactor is operating at full power (these PCMAs were discussed earlier, pp. 30–32). The report does make an indirect reference to this possibility when discussing transients in PWRs, where this type of PCMA may be more likely to occur because of the higher concentration of power output in a PWR core, namely, a higher core "power density." In referring to transients occurring in a PWR without reactor shutdown (for example, SCRAM), with all other safety systems working, the report remarks that the "core status" is "possible O.K.," which implies that core melting in an intact, pressurized reactor vessel cannot be ruled out.[58] The report does not elaborate.

Steam Explosion Probability

The Rasmussen Report asserts that a "steam explosion," that is, the explosive interaction of molten UO_2 with water coolant, has a "low probability" of occurring in a core meltdown accident.[59] Since the *steam explosion* is the chief mechanism postulated in this book for rupturing both the reactor vessel and the containment and for doing so *promptly* to possibly maximize the release of fission product radioactivity into the environment, it is necessary to address the Rasmussen Report's contention of a low probability of steam explosions, since it would seem to discount this author's analysis of the potential for accidents.

To begin with, the report expressly makes the contention of a low steam-explosion probability for only the LOCA-type core meltdown, in which molten UO_2 fuel would drop into a pool of unpressurized "saturated" (p. 37) water at the bottom of the reactor vessel or containment.[60] That is, the report's low steam-explosion probability claim does not apply to fuel melting by a PEA or PCMA, where the fuel melting conditions would be fundamentally different.

Furthermore, the contention in regard to LOCA-type core meltdowns is mostly speculation: First, virtually no experiments with molten UO_2 and water have been conducted, with the exception of an experiment in which tiny gram-size drops of molten UO_2 (less than 6/100 of a cubic inch) were dropped in water. This would not seem to be enough to produce an explosion, as a reactor will contain about 100 tons of UO_2. The Rasmussen Report admits that the gram-size tests are "by no means definitive":[61] "Virtually no data could be found which were directly applicable to the light-water-reactor meltdown situations considered in this program, i.e., large amounts of molten UO_2 falling into saturated or near-saturated water." [62]

Second, the gram-size experiments produced a "friable," solid UO_2 product that "readily" crumbled into a powder, which might indicate that the test was on the verge of generating a steam explosion, for it has been established that in order for a steam explosion to occur, the molten fuel, when interacting with water, will have to quickly disperse through the water as finely divided powder, so that the 5,000°F UO_2 can rapidly heat the water to explosive steam pressures. The grain size of the powder observed in the gram-size experiments is well within the range which the Rasmussen Report's theoretical model of steam explosions predicts will rupture both the reactor vessel and the containment.[63]

Third, an experiment with molten salt and water has produced violent steam explosions.[64] Here, a greater volume of molten material was involved (forty times more than in the gram-size UO_2 tests). Since molten salt and UO_2 have similar thermal properties, this experiment further emphasizes the possibility of steam explosion.

It appears that the only way to resolve the question of the probability of steam explosions in LOCA-type core meltdowns is to perform several full-scale, reactor-destruct-tests in order to ensure that all factors that can influence the outcome are included. More than one test would be needed to establish the probability of steam explosions, since it is known from experience of steam explosions caused by molten metal and water that a chance element is involved. However, no such tests, even using small-scale reactors, are planned, though the Rasmussen Report itself lists, among three "research suggestions," "large-scale" molten fuel-coolant interaction experiments.[65]

In regard to power excursion accidents (PEAs) which produce fuel melting, both the likelihood and the strength of steam explosions would presumably be much greater. In such power excursions, significant quantities of the fuel itself can be vaporized explosively, which could be a mechanism for dispersing molten fuel into the surrounding water coolant, thereby inducing steam explosions. In addition, the molten fuel would be generally hotter than a LOCA-type meltdown and is already near-intimately mixed with the water by virtue of the geometry of fuel rods immersed in water. (This is also the case for those PCMAs which produce fuel melting.) Historically, steam explosions in nuclear reactors were discovered in power excursion tests (the BORAX and SPERT tests mentioned earlier (p. 67) and in a PEA (ch. 4, n. 88), which involved metallic fuel, not oxide fuel). As with the LOCAs, large-scale PEA tests would be needed; though small-scale, SPERT-size oxide core tests would be very useful in this regard. In fact, however, *full-scale tests* would be necessary due to reactivity considerations, as previously discussed, in order to determine the severity of the power excursion, which will in turn determine the severity of the steam explosion that will surely result for excursions which produce fuel melting.

Probability of Accidents

The Rasmussen Report concludes overall that the chances of anyone being killed as the result of a reactor accident are, for example, 100,000 times less than the chances of being killed in a motor vehicle accident.[66] This conclusion is based on detailed probability analyses for the limited accidents considered in the report. Though the report's probability analysis is useful, such extremely low probability estimates for serious accidents are not reliable or absolute guides for assessing the risk of accidents, because of the *subjective* nature of the analysis. Yet, the NRC has argued that the estimates are completely *objective,* since they are based (in part) on actual failure rates observed on equipment and "bounding" probability values determined by expert psychological opinion as to the potential for human error.[67] Clearly, however, the estimates are based on a number of subjective judgments (assumptions). These subjectivities include:

1. Judgments made in selecting limited reactor operating periods from which failure data is taken, thereby excluding other incidents of failure (see p. 78).

2. Assumptions that equipment failure rates inferred from past data will hold for the future. For instance, the SCRAM switches that were once found totally inoperative had a prior history of no significant failure (see app. 2, no. 5).

3. Assumptions that all significant equipment failures or human errors that have actually occurred in nuclear plants have been reported. In this regard there is a question as to whether all nuclear power plants are required by regulation to report "occurrences or conditions that prevented or could have prevented a nuclear system from performing its safety function." According to the Nuclear Safety Information Center of Oak Ridge National Laboratory, such information "may" be required to be reported.[68]

4. Guesses of failure probability values for components and systems which only operating experience with a large number of reactors can confirm. For example, the probability of a control rod ejection reactivity accident (PEA) depends on the probability of a rupture of one of the pipes connecting the control rod drive mechanisms to the reactor vessel. Operation of the BWRs for their full design life would be required to determine the likelihood of rupture, since the failure due to stress fatigue or other loadings (perhaps abnormal loadings) is entirely dependent on the age of the plant and the manner in which it is actually operated. (After the forty-year life, can we expect a reactor refurbishing or stretch-out program that involves the continued use of the equipment?) Consider the opinion of the NRTS: "A realistic evaluation of the occurrence probability for the reactivity accident, however, cannot be made because the necessary statistical data are not and probably never will be available." [69]

5. Judgments about the potential for human carelessness or malpractice. The Rasmussen Report's "Human Reliability Analysis" is not based on data on human error associated with reactor abnormal incidents but on "substitute" data from other industries.[70] This use of substitute data ignores the *radiation factor,* which is peculiar to nuclear plants and which tends to inhibit quality maintenance; for example, a reactor mechanic or operator working in a radiation zone would tend to hurry his work, thereby inviting error. Moreover, the report assumes that all nuclear power plants will behave alike relative to the probability of failure occurrences—that is, the likelihood of an accident is determined from the probability per plant multiplied by the number of plants. However, though it may be assumed that the federal government's regulatory process will ensure that most nuclear plants will be operated carefully, there is always the possibility that one or two plants will be managed very poorly, to the point where many safety systems and repair work are neglected. There are a number of ways whereby such a situation could escape detection—bribery, for example.

6. Assumptions about component failures causing additional failures through unanticipated effects in such complex systems.

7. Assumptions regarding the course of physical processes during an accident for which adequate experimental data are lacking; for example, the report's low estimate of the probability of steam explosions given molten fuel contacting water.

The foregoing observations do not represent a full evaluation of the probability analysis of the Rasmussen Report. Due to the subjective nature of the analysis, it is felt that the first order of business should be to develop all essential objective information, such as all specific accident possibilities, scientific uncertainties of accident processes, experimental needs, and so forth. Eventually, of course, the probability of accidents will have to be thoroughly considered; and toward this end, the probability analysis of the report will have to be thoroughly scrutinized by independent authority, something which has not as yet been done.

Conclusion

Overall, the Rasmussen Report is unsatisfactory, as many crucial assumptions and assertions are made without specific rationale and supporting references. Indeed, the report contains a notice which disclaims any "responsibility for the accuracy, completeness, or usefulness of any information" in the report; no such disclaimer is made in the WASH-740 report. Moreover, the author or authors of the volume of the Rasmussen Report on "Accident Consequences" are unnamed;[71] in contrast, the WASH-740 report contains the authors' names. The acceptance of responsibility and the acknowledgment of authorship would seem to be wholesome, traditional ways of allowing society to judge the validity of health and safety assessments through knowledge of the qualifications, reputations, and vested interests of the authors of such assessments. In summary, the Rasmussen Report is no reliable analysis on which to base a conclusion that reactors are safe, as the nuclear community has done. On the other hand, the report is a valuable and needed contribution to the national inquiry into nuclear reactor safety, as it has helped greatly to develop information about reactor accident hazards and safeguards; as such, it serves a useful basis for discussion.

Seven
The 1964–1965 Reexamination of WASH-740

IN 1964–65, the AEC and the Brookhaven National Laboratory undertook to revise the 1957 WASH-740 accident analysis in light of the fact that much larger plants were just then being approved for construction by the AEC. However, a report of the study was never issued. In 1973 the AEC made public the papers generated by the 1964–65 study, including draft reports, memoranda, meeting minutes, and so forth. Neither the AEC (NRC) nor Brookhaven has vouched for the papers, however, and so they are inconclusive. Even so, inasmuch as the papers are a matter of public record [1] and are obviously relevant, it would be appropriate to mention some of the highlights relative to the issues of this book, provided the above qualification of inconclusiveness is kept in mind.

1. The initial finding of the study was that a land area of the size of Pennsylvania could be seriously contaminated ("disaster") due to a reactor accident.[2] However, later drafts of the report removed any quantitative estimates of the accident consequences. In a draft letter to the AEC, the Brookhaven Laboraory stated: "We have therefore believed it most useful to present the results of the survey in qualitative form rather than quantitative. We believe that the qualitative conclusions have greater validity than any numerical ones that we could furnish." [3]

2. Later drafts of the report assume 1% release of Sr 90 upon a core meltdown accident, based on the previously discussed Browning experiments at Oak Ridge Laboratory, which are relied on by the Rasmussen Report for its low 6% assumption

(see pp. 82–83). However, the associate director of the Oak Ridge Laboratory and others involved in the 1964–65 study cautioned that the Browning experiments could not be extrapolated to the large reactor situation and that the percentage release could be greater (vastly different surface-to-volume ratios were cited).[4]

3. As for the particle size of released fission products, the study concluded that the size would be small (equal to or less than one micron diameter), which conflicts with the 125 times heavier particle assumed in the Rasmussen Report.[5] Recall that the smaller, one-micron-size particle, assumed in WASH-740, would allow greater-size land area contamination (see p. 84).

4. As for the ground contamination limit for strontium 90, the 1964–65 study used basically the same value assumed in the Rasmussen Report, instead of the more stringent WASH-740 value. However, the study's theoretical derivation of the relaxed contamination limit differs with a published analysis by a factor which would explain the less stringent contamination limit[6] (the factor accounts for the rate of uptake of Sr 90 by plants and cows). Also, there is an indication that the limit might be based on relaxing the legal maximum permissible levels (MPL) of radiation exposure to the general population, which may be why the report was never issued, since MPLs have, if anything, been tightened with the passage of time. Clearly, what is needed is a thorough, plain derivation of the radioactivity contamination limits that is fully justified. Such a derivation would have to show how the more stringent WASH-740 limit, and the justification for it in the WASH-740 report, can be relaxed, if a relaxation can be defended. Furthermore, this derivation should be checked by plant and soil and health authorities in the various state governments to reach some agreed-upon contamination limit for use in reactor accident risk evaluations. In view of the strong basis for questioning the Rasmussen Report's contamination limits for Sr 90, and its treatment of accident consequences generally, it is recommended that the report's estimates of the genetic effects of reactor accidents—that is, effects on future generations of humans—be reviewed thoroughly as well, with respect to contamination limits for radioactivity.

5. A draft report among the papers considered a hypothetical case in which all of the "airborne fission products" were released into the atmosphere at the same time as the coolant (turned steam). It was found that the heat of the mixture would carry the radioactivity "into the upper atmosphere, where it would be harmlessly dispersed." [7] This is very interesting indeed, but it may not be realistic; the fission product release may follow the release of

coolant, even for the most severe PEAs. And was the wash-out effect of rain considered?

In the end, no report of the 1964–65 study was issued, though "separate reports" were promised by the AEC chairman in a June 18, 1965, letter to Congress, which never materialized.[8] Meanwhile, construction permits for over 109 large reactors have been approved since 1965, without a rigorous, fully justified, hazards analysis which scientists would stand behind. The need for such still exists.

Eight
The American Physical Society Report

THE FEDERAL government has sponsored one other widely heralded reactor safety study, conducted by a study group of the American Physical Society (APS) under a $135,000 grant from the AEC and the National Science Foundation.[1] According to the APS report, the study group has "not uncovered reasons for substantial short-range concern regarding risk of accidents in light-water reactors."[2] However, as with the Rasmussen Report, the APS study, while very useful, does not adequately treat the safety issue.

To begin with, the APS report treats only the LOCA and related ECCS technical matters, and spontaneous reactor vessel rupture. That is, *the APS report neglects the PEAs altogether,* which should qualify the APS conclusion. Furthermore, the report treats superficially the subject of transients (for example, it merely mentions some of the PCMAs, specifically, heat exchange accidents), and thus provides no grounds for drawing any conclusions regarding the risks of this class of accidents.

In regard to the failure resistance of fuel rods, including crumbling, which is crucial to the safety of the DBAs, the APS report concludes: "A great deal of research . . . has gone into guaranteeing the integrity of the fuel elements and cladding . . . and the ability to control the reactor under both normal and abnormal conditions."[3] However, this conflicts with the facts. Recall that the National Reactor Testing Station (NRTS), reporting the results of some experiments on the effects of fuel burn-up on fuel rod failure thresholds for power excursions, stated that "unexpected" low thresholds were observed which, the report added, "cast serious questions" upon reactor. safety with "crucial" implications. The NRTS "concluded that additional tests are urgently needed . . . for power reactor safety issues," [4] but, instead, the fuel test program was abruptly terminated in 1970–71. The APS report does not address these NRTS documents. Fuel testing under accident conditions is just now being resumed in the Power Burst Facility (limited PCMA testing), with PEA- and LOCA-related testing in PBF still quite a ways off. Therefore, it cannot be said that a great deal of research has been performed; and certainly there is no guarantee of safe fuel performance under a DBA.

Somewhat contradictorily, the APS report vaguely concludes that "transient accidents remain a serious concern" [5] and states without elaboration that "[s]ome fuel and cladding problems still need careful evaluation." [6] These statements leave the public to wonder what each means and what their significance is.

In regard to experiments, the APS report asserts that the Reactor Safety Research Program of the AEC is based partly on:

> Integral system experiments, smaller in scale than a full LWR [light water reactor], but hopefully large enough to test the ability of the [computer] codes to predict system effects accurately under a range of LOCA and *transient accident* conditions.[7]

This statement is not true, if by "transient accident" the APS meant to include power excursion accidents,* for no such tests for PEAs are planned, or possible, with existing facilities. The APS report based its statement on an article by the AEC's Reactor Safety Research director in *Nuclear Safety,* in which the PPF testing was characterized incorrectly as an "integral experiment." [8] The PBF in no way can integrally verify the DB-PEA theory (see pp. 72–73).

A 1963 NRTS report on the PBF design[9] emphasizes that the PBF is designed only for " 'differential' or subassembly tests," in which processes are "studied separately," and distinguishes such testing from "integral-core" testing. Concerning integral-core testing, the report asserted: "It is evident that continued studies of this type are necessary" as well. Indeed, a year later, in 1964,

* The APS report links the PBF testing with the "integral" testing under "transient accident conditions" (pp. S53–59).

the NRTS laboratory, the designers of PBF, issued the previously mentioned PTR-738 "internal report," in which a number of "integral-core tests" were recommended, including full-scale destructive tests. The PBF testing was not placed in this integral test category in the PTR-738 report, but in the category of "In-Pile Differential Experiments." [10]

Returning to the PBF design report, the purpose of PBF subassembly testing was to complement the integral testing, by providing data that could not be measured in the integral-core tests, since to attempt to measure detailed effects in the integral tests would affect their results. Hence, the PBF differential tests and the integral-core tests were *both* considered necessary "for obtaining a fundamental understanding" of power excursions.[11] It should be clear that by now characterizing the PBF tests as "integral" experiments, instead of "differential," the AEC incorrectly conveys to the public a certain sufficiency in the PBF testing coming up. In misinterpreting the character of the PBF testing, the APS report is an example of how the public is being misled by the AEC's careless use of terms.

Regarding the detailed evaluation of the LOCA in the APS report, it is unsettling that the report contains numerous important assertions which tend to reduce the concern for the LOCA hazards without citing any supporting references, an omission which casts further doubt on the validity of the report's overall conclusion. For example, the report merely asserts that the "hydraulic forces" on the fuel in a LOCA are not significant; that containment-rupturing steam explosions require a "large fraction" of molten core to effect them; and that the maximum experimentally observed Sr release fraction from UO_2 was only about 10%; but no references for these statements are provided.[12]

Despite its shortcomings, the APS report is still very useful, since it does focus on a great many LOCA and ECCS issues. The report's appraisals, though largely unsupported by references, are authoritative, since at least two key scientists in the reactor safety test program, S. Johnson and G. Brockett,[13] were heavy consultants to the APS group. A primary example of their influence is the report's questioning the adequacy of LOFT to verify DB-LOCA calculational methods, because of uncertainty in the "scaling" factors—that is, LOFT, being small-scale, cannot reproduce many large-scale effects.[14] Indeed, the APS report is replete with serious questions about the reliability of the ECCS to control LOCAs; and yet, the report concludes that there are no reasons for short-term concern about accident risks, which is baffling, since a LOCA (that is, a sudden rupture of a reactor coolant pipe) can happen at any time, as in the boiler-water pipe rupture at the Indian Point 2 PWR.[15]

As argued before, an adequate experimental verification program for the DB-LOCA and associated ECCS performance may be practical, though it necessarily would require several full-scale reactor DB-LOCA tests, using two test reactors (see p. 64). Aside from the question of irradiated fuel performance in a LOCA and full-scale three-dimensional effects in a PWR, the ECCS should function satisfactorily in a DB-LOCA. Thus, the narrowly defined DB-LOCA is a technical issue that could be raised and resolved experimentally without any large risk of experimental failure to alarm the public while, incidentally, providing the laboratories and universities with lucrative research projects in the process. To raise the PEA and PCMA issues and discuss their significance and implications, however, would fundamentally question the use of nuclear power, which may explain the reluctance of the nuclear engineering and physics community to discuss them.

Incidentally, the APS report contains some estimates of the damage consequences of a reactor accident, which are roughly the same as those of the Rasmussen Report (WASH-1400), except that the APS group did not calculate the "acute deaths" caused by direct exposure to the accident-caused radioactive cloud. Though the APS group calculated higher long-term cancer effects and genetic defects for the worst accident treated (10,000 cancer deaths *v.* 310 in the Rasmussen Report, due primarily to the alleged neglect in the Rasmussen Report of the long-term ground contamination of Cs 137),[16] the APS report's overall damage estimates are minor compared to this author's WASH-740-based extrapolations (long-term cancer deaths were not addressed in the WASH-740 report). However, the APS estimates are based essentially on the same assumptions of the Rasmussen Report; a low Sr 90 release fraction, heavy radioactive dust particles, twentyfold higher Sr 90 ground contamination limit, and so on.[17]

Moreover, the APS report disclaimed any accuracy or completeness for its estimates:

> We have not discussed here all the uncertainties in the calculation of the consequences of the Draft WASH-1400 reference accident. It would appear that more effort is needed in the exploration-of-consequences calculations before we can be certain that all important effects have been considered.
>
> Because the art of consequence modeling appears to be at such a rudimentary level of development, our results should not be regarded as definitive.[18]

As for contamination limits, APS stated:

> The thresholds are so arbitrary and the possibilities for land decontamination or reassignment so unevaluated, however, that we cannot here provide a definitive estimate.[19]

These disclaimers, plus the facts that acute deaths were not calculated in the APS report and that the assumptions underlying the Rasmussen Report's accident consequences estimates have not been supported by reasoned analysis or documentation, mean that the WASH-740 report has not been superseded by the Rasmussen and APS reports. Therefore, the WASH-740 report remains the only authoritative and reasoned analysis from which to estimate the disaster potential of a reactor accident; hence, the WASH-740 extrapolations given at the outset of this synopsis are as valid as any other estimates of accident consequences offered so far.

In conclusion, the APS report avoids surveying the entire reactor accident hazard and questioning fundamentally the safety of nuclear power. Instead, the thrust of the report seems to be toward finding design measures and gathering theoretical and experimental data for possibly lessening the risk and level of disaster, but only in regard to the LOCA; that is, without regard to whether the risks of PCMAs and PEAs are overwhelming.[20] In all, the APS report is no basis for a conclusion that reactors are safe, or can be made safe.

Nine
The Secret Hazards Research Recommendations of 1964: A Critical Juncture

AN INTERNAL National Reactor Testing Station report (PTR-738) was mentioned earlier in connection with the need for full-scale tests to verify the calculational theory for design basis, power excursion accidents (DB-PEAs) and to investigate the PEA potential for worse possible accidents. This extremely important NRTS report was issued in late 1964 or January 1965 at a crucial juncture in nuclear power development. Before that time about fifteen small reactors of varying concepts had been tested (average size: 75 megawatts, including a 163-Mw PWR, a 220-Mw BWR, and a 175-Mw PWR). From this reactor experience emerged two designs, the PWR and BWR, which demonstrated economic potential, provided the reactor sizes were greatly increased to take advantage of economies of scale. The industry then designed large-size BWRs and PWRs.[1] At that point the NRTS laboratory conducted a review of PEAs in *large* PWRs and BWRs, the results of which were presented in the NRTS PTR-738 report, which is quite comprehensive. The report concluded: (1) The excess reactivity of the PWR and BWR cores is large enough that PEAs could be "catastrophic." This conclusion covered "single-failure reactivity insertions," such as a control-rod dropout, as well as multiple-failure PEAs, as the report noted the possibility of unforeseen autocatalytic reactivity effects occurring in single-failure reactivity accidents. The DB-PEAs are single-failure reactivity "insertions," that is, reactivity increases. (2) Many autocatalytic reactivity effects are conceivable. (3) Complicating neutron dynamic effects arise with large cores. And (4) an intensive, six-year minimum theoretical and experimental research program is necessary to investigate these matters. The recommended research program included excursion testing of individual fuel rods and fuel rod bundles, including testing in the Power Burst Facility, which was then being designed; large-core neutron dynamics experiments; and "Destructive Tests Involving Essentially Full-Scale Operating Power Reactor Systems."[2] (The PTR-738 report distinguishes the PBF testing from "integral core" testing in which a whole core is tested.) Regarding the full-scale destructive tests, the NRTS report concluded that "The program cannot be considered complete until this type of test is done."[3] The report added:

> In submitting the following program recommendations, it is considered that time is of the essence, not only because of the large number of present and proposed power reactors of the type considered in this review, but because of the inherent amount of time required to complete a program of this magnitude.[4]

Unfortunately, the PTR-738 report was never published or referenced in the literature, nor was it revealed until 1974, ten years after it was prepared.[5] In effect, it was kept secret. Moreover, no other quantitative theoretical analysis of the PEA potential of PWRs and BWRs has ever been published * or referenced in the literature. Recall that the 1966 detailed technical proposal of the NRTS for the large-core dynamics experimental program, which reiterated the need for large-core testing, was also classified "internal report" and held the "PTR" serial heading. It too had been kept secret (see pp. 48–49). (The NRC should publish a list of all

* Except an obscure 1970 European report (*Water Cooled Reactor Safety*, Committee on Reactor Safety Technology, European Nuclear Energy Agency, May 1970, chap. 9), which barely treats the subject. This report presents some theoretical predictions of only two severe PEAs in BWRs (control rod ejection and control rod dropout), though they were far from the worst possible from the standpoint of *reactivity worth* of the control rods (only 2.5% was assumed). Also, in 1969 some calculations of the power excursion potential of the small SPERT-III test reactor were published, which indicated the potential for catastrophic explosions for that reactor; but the relevance and implications for the PEA potential of large commercial PWRs and BWRs were not explored. (IDO-17281, pp. 69–75; and J. E. Houghtaling and J. E. Grund, "Analytical and Experimental Investigations of the Kinetic Behavior of a Uranium-Dioxide-Fueled Pressurized-Water Reactor," *Nuclear Science and Engineering* 36:412, 424–25.)

of the PTR-series reports of the NRTS and make the reports available to the public.)

Instead of postponing nuclear development until the recommended research was performed, so that safety judgments could be made "on a properly objective basis," [6] the AEC suppressed the report and embarked at that point on a massive construction program of *large* BWRs and PWRs with the issuance in 1964 of construction permits for the first large plants—the 430-Mw San Onofre PWR, the 575-Mw Haddam Neck PWR, and the 640-Mw Oyster Creek BWR. The AEC has since approved over 109 construction permits for even larger reactors (up to 1,250-Mw).[7] Thus, the AEC rejected its expert scientific advice and proceeded, borrowing from the NRTS report, "in the face of little or no factual information on areas of importance." [8] (Since that time, only small pieces of the program have been attempted; and then, when the data indicated alarm, as in the single-rod, irradiated fuel tests and in the simple, subcritical, large-core neutron dynamic experiments, their respective programs were terminated.)[9] In contrast, the AEC assured the public in 1964 in a "public information" booklet on *Atomic Power Safety* that:

> The safety of atomic power is studied as well as practiced. The U.S. Atomic Energy Commission is sponsoring a major research and test program in this field. . . .
>
> Through safety research and tests, the atomic power industry is continuously strengthening the most important safeguard any industry has—namely, knowledge of the causes and consequences of accidents and of the dependability of safeguards. Learning about accidents before they occur is part of the basic fabric of the safety of atomic power.[10]

It is clear that the public was misled.

The NRC was recently asked for the documentation of "the AEC's technical basis for not approving the research program that was recommended in the PTR-738 report." [11] The NRC's director of research, Dr. H. Kouts, replied that the "complete AEC files for the pertinent period are not available." [12] Instead of disclosing whatever AEC documents do exist, if any, Dr. Kouts "reconstruct[ed]" the AEC's reasons "from personal memory of individuals who were peripherally involved." Dr. Kouts concluded that "core destructive tests" were considered and that a design of a "Nuclear Test Facility" (NTF) was made for that purpose.* However, said Dr. Kouts, the AEC felt that there existed insufficient

* The documents related to this NTF should be disclosed.

"basic" data of fuel rod failure characteristics "to constitute a rational basis for planning and predicting core destructive tests"; for this reason, the PBF-type testing "was given priority over the NTF and other proposals for conducting core destructive tests."

This reconstructed rationale does not justify the authorization of over 109 large reactors and the rejection of the recommended NRTS experimental program and schedule. Firstly, the PTR-738 report recognized the need for doing the PBF testing first, to obtain the basic data necessary for the "formulation of parameters for full-scale destructive tests." That is, the PTR-738 report also gave priority to the PBF testing, but in the sense that full-scale testing was definitely to follow. Secondly, Dr. Kouts' reply admits that even the basic data has not yet been obtained: "The Power Burst Facility has now been constructed and is actively used in experiments to supply the basic data that are desired." Clearly, the NRC's reconstructed rationale does *not* conflict with the PTR-738 report; rather it is in every respect consistent with the rationale for the recommended experimental program given in PTR-738. Furthermore, the NRC's reasons do not explain why the recommended experiments beyond the PBF testing are not included in the NRC's *Reactor Safety Research Program.*[13]

It appears, therefore, that the AEC has simply slowed the schedule of the nonintegral experiments and rejected without scientific grounds the recommended large-core kinetics experiments and the full-scale tests. The major parts of the PBF testing were to be finished by 1970 and the rest of the PTR-738 program by 1972, according to the PTR-738 plan.

One is justified in suspecting, therefore, that the AEC simply did not want to carry out the PTR-738 program for fear of finding that the DB-PEAs are uncontrollable or that the accident potential is otherwise so large that the public would not tolerate the risk. In any event, the course the AEC has taken is patently reckless. In the opinion of this author, there was no real need to begin building the large reactors in 1964. The nation could have pursued six to ten years of testing without any economic dislocations, such as the closing down of nuclear plants, other than three small ones. Indeed, many jobs and contracts would have been awarded in connection with the recommended research, which would at that time have given a boost to the economy in some respects. Hence, by disregarding those recommendations, it would appear that the AEC acted to enrich the special interests they served, instead of ensuring the public health and safety against reactor accidents with verified accident theory and full hazards investigation.

Since the PTR-738 report, design changes have at least been

made to reduce the probability of severe PEA,[14] which led to the present "design-basis accidents," but this does not eliminate the need for experimental verification of the DB-PEA theory. Nor does it address the question of the accident potential of the worst power excursion situation; again, the probability factor is subjective.

As a result of the suppression or disregard of scientific analysis and experimental results, most, if not all, of the nuclear engineers are engaged in building and operating machines which they do not understand relative to accidents, though they may think they do. To illustrate, the following persons of the nuclear industry admitted publicly that they were not aware of the NRTS' report PTR-738[15]: Professor Norman Rasmussen, the chairman of the AEC's *Reactor Safety Study;* Bernard Fox, project engineer of the two 1,150-Mw Montague BWRs for Northeast Utilities; Dr. James Coughlin, nuclear engineer and Nuclear Energy Vice-President of Public Service, Indiana; nuclear safety specialist, Professor Robert E. Bailey of Purdue University; and Dr. Andrew Kadak, who is a nuclear engineer in a reactor manufacturing company and coauthor of a favorable report on nuclear reactor safety, which did not even mention the PEAs (or PCMAs).[16] Dr. Kouts, the NRC's director of research, was unaware of the "complete failure" of the DB-PEA theory, WIGLE, to predict large-core kinetics experiments;[17] and the NRC's manager of safety research on fuel rods was not aware of the test report on multirod testing in which fuel crumbling occurred.[18]

The question has been raised whether the theory of power excursions has advanced enough since 1964 so that at least credible estimates can be made of the upper limits of the PEA potential, which may hopefully be acceptable. To answer this, there are indeed more advanced theories for calculating power excursions since the 1964 calculations in the PTR-738 report, but they have *not* been applied to the excursion accidents calculated in the PTR-738 report—that is, there is no updated report.[19] Moreover, there is a strong indication that the more advanced theories predict *worse* excursions, not lesser ones. (See pp. 44, 77 about the effect of the "more refined calculations.") A major point of this book is that the latest theories have not been applied to the worse possible accidents (WPAs)—those more severe than the Design Basis Accidents (DBAs)—to predict the full accident potential with or without autocatalytic reactivity effects. There are conceivable disastrous excursions without the need for autocatalytic reactivity effects—for example, the straightforward reactivity rise potentials of a control rod ejection accident—which could be calculated (see pp. 24–26). Although using the advanced theories to calculate the reactivity effects of potential or conceivable motions of core materials, to investigate autocatalytic reactivity possibilities for causing severe nuclear runaway, is probably beyond computer capability, such calculations ought to be attempted, as some insight could be achieved, at least.

To better appreciate the difficulty of using advanced theory, consider that while a one-dimensional power excursion calculation of just the neutronics (excluding the costs of calculating the core materials motion) costs $4 per accident calculation, a two-dimensional calculation, which increases the number of variables by one, costs $2,000 per accident calculation.[20] "One- or two-dimensional" theory means that the spatial variations of the fissioning in the core are assumed to vary with time only over one or two dimensions, which is not realistic. To be realistic, especially to investigate autocatalysis, *three* dimensions are necessary to explore the real world, which by extrapolation would mean roughly a million dollars per calculation (?).* In addition, the theory being used is known as "diffusion theory," which assumes that the neutrons fly around nearly equally in all directions; but when holes form or gross core deformation occurs, diffusion theory would not be reliable. "Transport theory" would then be required, which adds two more variables to keep track of neutron directions. Further, if the neutron speeds cannot be divided into two groups (fast and slow) but must be divided into more groups, to account for effects of the transient and gross core deformations on the neutron speeds, the practicality of the calculation becomes even more remote. There is also the problem of calculating the complex fuel and coolant motion, which causes reactivity changes.

In the early days when cores were tiny, such as in the SPERT reactors, power excursions could well be predicted with zero-dimensional diffusion theory, but with today's cores, the advanced theory predicts very complicating large-core effects; and additional autocatalytic effects arise conceptually as well. That is why the PTR-738 report recommended the large-core kinetics tests and full-scale excursion destructive testing; and why advanced theories had to be developed, which may very well be impractical to apply to the WPAs.

Therefore, it may very well be impossible to calculate a

* Recall that the Savannah River Laboratory's calculations of the neutron wave experiment suggest that 3-D calculations may be necessary in any event (see p. 49).

credible upper bound of the power excursion potential of PWRs and BWRs, or even to calculate credibly the more likely of severe PEAs, or to calculate credibly the course of Rasmussen Report's "transients" which produce fuel melting while the core is still critical. If this is so, then one will be forced to calculate definite upper limits or bounds of the explosion potential of the various accident possibilities on the basis of conceivable processes which cannot be shown to be impossible.

Assuming, however, that theories are proposed by the nuclear community which are "judged" to be adequate approximations of rigorous theory, and which predict an acceptable level of severity for those accidents which are more than remotely possible, they would still have to be *verified experimentally*—a requirement which has not even been satisfied for the DBAs.

To be sure, the present DBA theory is partially founded on experiment, but where it is not, the industry is relying on theoretical "models" and mathematical approximations which still leave the whole theory unverified. Consider the opinion of the reactor dynamics expert at MIT, Professor E. P. Gyftopoulos:

> Any physical concepts or mathematical models or any analytical or computer results, regardless of their degree of elegance and sophistication, are of no consequence, particularly in questions of nuclear safety, if they have not been or cannot be justified and repeatedly verified by experiment.[21]

The Newtonian Experimental Philosophy

It ought not have to be argued that unverified theoretical predictions of the course of reactor accidents should not be relied on to conclude that nuclear reactors are safe. The history of science is full of instances where theory has been disproven by experiment and observation. To be sure, many such instances have occurred in reactor research. Examples are: the unpredicted reactor explosion in a SPERT-1 power excursion test;[22] the "somewhat unexpected destructiveness" of the BORAX-I excursion;[23] the autocatalytic over-power transient that caused the EBR-I total core meltdown;[24] the failure of the WIGLE neutron kinetics theory in a large-core neutron dynamics experiment; and the failure of a theoretical calculation to predict the results of a LOCA-ECCS simulation experiment (the "semi-scale tests").[25] (In the semi-scale test, the simulated ECCS for a PWR failed to fill the test vessel to cool the artificially heated, simulated reactor core; this was evidently not predicted.)[26]

Unverified reactor theory, despite the care that some may think goes into its development, is merely *hypothesis* and *conjecture.* By relying on such for safety calculations, the nuclear community is discarding the time-proven *experimental philosophy* of science, thus venturing dangerously into the unknown. In view of the reactor hazards, we would do well to review this philosophy, which is perhaps best enunciated in the preface by Mr. Roger Cotes to Isaac Newton's *The Mathematical Principle of Natural Philosophy,* the classic treatise in which the laws of motion were presented and proven on experimental grounds:

> [W]e must not seek from uncertain conjectures, but learn [the laws of Nature] from observations and experiments. He who thinks to find the true principles of physics and the laws of natural things by the force alone of his own mind, and the internal light of his reason; must either suppose that the World exists by necessity, and by the same necessity follows the laws proposed; or if the order of Nature was established by the will of God, that himself, a miserable reptile, can tell what was fittest to be done. All sound and true philosophy is founded on the appearances of things; . . .[27]

Newton put it another way:

> For whatever is not deduc'd from the phenomena,* is to be called an hypothesis; and hypotheses, whether metaphysical or physical, whether of occult qualities or mechanical, have no place in experimental philosophy. In this philosophy particular propositions are inferr'd from the phenomena, and afterwards render'd general by induction.[28]

Of course, past reactor safety research has included many experiments which have investigated isolated phenomena, mostly on a small scale; but such experiments are not sufficient to test large-scale interactions of physical processes. Also, small-scale tests do not include many crucial processes or effects that occur or are predicted to occur during acidents in large-scale reactors. Indeed, in many important instances even small-scale reactor tests have yet to be performed; for example, power excursions which produce fuel melting, power excursions with irradiated fuel, integrated LOCA-ECCS tests, and power-cooling mismatch tests.

* The term *phenomenon* used by Newton means, according to Webster, "any fact, circumstance, or experience that is apparent to the senses and that can be scientifically described." The term has its origin from the Greek word *phainomenon,* "to appear." Therefore, by *phenomena,* Newton meant *experimental facts.*

Moreover, the very few large-core experiments which have been performed, namely, the neutron dynamics experiments, did not include power excursions and associated reactivity feedback effects.

However, the essential point of the Newtonian experimental philosophy as it applies to reactor accidents is not that *full-scale reactor tests* are necessary to investigate predicted or conceivable effects and their interaction, though this is true, but that such tests are necessary in order to *learn* how full-scale reactors will respond to malfunction. For it is clearly possible that *unforeseen* phenomena will occur—that is, no matter how we might theorize on the course of assumed accidents, we *really* will not know the accident potential unless we perform experiments or otherwise experience the accidents. The task of physics is to construct a theory which describes natural phenomena, such as a reactor accident—or a test of one—that has actually occurred, and not to develop a theory of how we might suppose a real reactor might behave based solely on observing *other* phenomena, such as those observed in small-scale or isolated-effects experiments. A successful theory—one that is verified by a series of full-scale tests—would allow us to gain more knowledge of reactor accidents than simple observation of the test results. That is, the usefulness of theory, when verified and rendered general by induction, is that it would make additional, costly, full-scale tests unnecessary for calculating the course of other assumed accidents in which the circumstances, such as the reactor design, vary slightly from, but are still within the bounds of, those of the verification experiments.

This point deserves dwelling upon, since it addresses the fundamental difference in philosophy at the heart of the reactor safety issue. The AEC emphatically ascribes to the *hypothetical philosophy* as opposed to the Newtonian *experimental philosophy* and has stated that "The concept of attempting to determine the maximum accident potential by direct core destructive tests has repeatedly been determined to be a nonproductive and unnecessary way of obtaining a responsible understanding of phenomena."* [29]

As concluded earlier, full-scale tests in regard to the DBAs may be practical, though this is doubtful. However, such tests in

* Here, the AEC misuses the term *phenomena*, which strictly signifies any fact or experience that has been observed or measured experimentally. Since large-core reactor accidents have not been observed (e.g., PEAs), there can be no "responsible understanding" of it. This is the signification of *phenomena* used by Newton.

regard to WPAs are definitely not practical, except perhaps for several maximum-conceivable-accident-type tests (but then, maximum-accident tests would probably be too hazardous). Judgments about the practicality of full-scale tests probably have more to do with the formation of the AEC's research philosophy than anything else, except, cynically, a possible desire to avoid crucial tests of safety claims. Just because experiments are impractical, however, does not negate their necessity.

Incidentally, this author is not at this time recommending that full-scale tests be conducted, since such tests may be too hazardous in themselves. Rather, his point is that no prediction of the course of assumed accidents ought to be relied on for assessing the public health and safety and deciding that reactors are safe, unless the theory behind the predictions is demonstrated by full-scale tests. As for a decision on whether to conduct any full-scale tests, this should be deferred until a thorough, objective, scientific review of the matter is performed, and society has had a chance to consider the safety of nuclear power plants with the aid of a full hazards analysis.

Even in regard to experimentation on small-scale reactors and isolated effects, there exists an admitted gross deficiency of data for "verifying" many aspects of reactor accident calculations. This can be seen by examining the Nuclear Regulatory Commission's recently issued report on its *Reactor Safety Research Program.*[30] Many of the experiments now being planned were recommended as far back as 1964, and again in 1970,[31] also, much of the experimental data regarding DBAs from this research program will not be obtained until 1978 at least.[32] Despite these admitted shortcomings, the AEC has authorized, and the NRC continues to authorize, the construction and operation of reactors.

This completes the analysis of the accident hazards of water-cooled reactors. We defer a conclusion, however, until we analyze the explosion hazard of the advanced, "breeder" reactor, because, as we shall see, a judgment on the safety of the present nuclear power plants is tied to the breeder reactor safety issue.

Ten
The Explosion Hazard of the Advanced "Breeder" Reactor (LMFBR)

THE LIQUID metal-cooled, fast neutron, breeder reactor (LMFBR) is an entirely different power reactor concept than the water-cooled reactor. It is designed especially to produce or "breed" fissionable material as a by-product, namely, plutonium fuel, by certain nuclear reactions in the reactor that convert plentiful uranium-238, a weakly fissionable species of uranium, into plutonium. The objective of the LMFBR is to produce more plutonium than is consumed in an operating cycle (about 7% more per year). The excess fuel can then be used to start up other LMFBRs and to fuel the PWRs and BWRs when the useful reserves of rare fissionable uranium (U-235) are depleted, which is estimated to occur in about thirty years.[1] Hence, the PWRs and BWRs will ultimately depend on the LMFBR. (The reserves and present stockpiles of U-238 would last the U.S. for a thousand years or so, using the combination of LMFBRs and the water reactors.)

The AEC has projected that about one thousand large LMFBRs would eventually be built, along with a like number of water-cooled reactors.[2] A license application to build the "LMFBR Demonstration Plant" in Tennessee is presently pending; and a smaller LMFBR-like reactor, called the Fast Flux Test Facility (FFTF) is under construction, which is to be used to test LMFBR fuels under actual LMFBR core conditions encountered in normal operation. The FFTF differs from an LMFBR in that the plutonium-fueled core is not surrounded by uranium 238 for breeding purposes; otherwise, it is basically the same as an LMFBR, from the standpoint of accident hazards.

Unfortunately, the LMFBR has a power excursion (nuclear

runaway) potential—indeed, a potential for *nuclear explosion* as distinguished from a *steam explosion*—which is even more serious than that of the water-cooled reactors.[3] This LMFBR explosion potential has extremely grave implications, especially because such a nuclear explosion would produce radioactive plutonium dust, which is extremely toxic. A steam explosion, or more accurately, a "coolant vapor explosion," is defined as the explosive vaporization (boiling) of coolant due to the contact of the coolant with extremely hot, molten fuel. This appears to be the only possible mode of explosion in the water-cooled reactors. A nuclear explosion, on the other hand, is defined as an explosive vaporization of the fuel itself, which involves higher temperatures and potentially a much stronger explosion than a coolant vapor explosion. An LMFBR could also produce strong coolant vapor explosions upon core melting, which could add to the explosion force or by themselves be dangerous. As will be discussed, nuclear explosions of the order of 20,000 pounds TNT-equivalent are theoretically conceivable. For comparison, the maximum economical containment capability is about 1,000 pounds TNT.[4] Thus, such an explosive power excursion would vaporize the entire core, rupture the containment, and blow, say, half of the core, amounting to tons of radioactive plutonium, into the atmosphere —and boil off practically all of the fission products and blow them into the atmosphere as well. Moreover, the core vaporization process presumably would disperse the plutonium and fission products in the form of a superfine dust, causing severe, geographically widespread, ground contamination.

The consequences of such a heavy fission product release were estimated previously in chapter 1. The additional consequences due to plutonium release could be even worse. Because of the extreme toxicity of plutonium, its release (assuming two tons)* and its long life (up to 600 to 24,000 years "half-life," depending on the isotopic species of plutonium) could cause permanent abandonment of over 150,000 square miles (an area the size of Illinois, Indiana, Ohio, and half of Pennsylvania combined). This estimate is based on simply substituting plutonium for the fission products in the atmospheric dispersal-fallout calculation of the WASH-740 report (that is, no extrapolations) [5] and finding that the ground contamination level at the boundary of the 150,000-square-mile zone exceeds a proposed contamination limit of one microgram of plutonium 239 isotope per square meter.[6] Actually,

* From a core containing up to three tons of plutonium, as contemplated for planned 1,000-Mw LMFBRs, to seven tons for larger projected LMFBRs (WASH-1184, pp. 34–40).

an LMFBR core will contain other, more radioactive, isotopes of plutonium, such that there will be about three to eight times more Pu 239—equivalent radioactivity than if the core were 100% Pu 239.[7] Hence, the equivalent Pu 239 contamination of the 150,000-square-mile zone would be three to eight micrograms/m^2 at the zone boundary, and greater inside the zone and closer to the accident. This level exceeds the one microgram/m^2 level proposed as a contamination limit by Willrich and Taylor, who stated that any ground contaminated above one microgram/m^2 "would be likely to be deemed unacceptable for public health," [8] and also exceeds the Rasmussen Report's contamination limit of three micrograms/m^2, above which "relocation" is to be required.[9]

Incidentally, the above use of the WASH-740 calculation does not involve those aspects of the WASH-740 analysis which differ with the Rasmussen Report; the differences appear to arise in the assumptions of the radioactive fallout dust particle size and contamination limits for Sr 90 following a meltdown accident, and not in the atmospheric dispersal-fallout aspects, were the dust particle size the same between the two reports. This assumes that a severe nuclear explosion of an LMFBR core would generate dust particles of one micron size or less, which is the basis for the 150,000-square-mile ground contamination value in WASH-740. In view of the extreme temperature of such a nuclear explosion, the assumption seems appropriate.

If the "hot particle" theory for lung cancer induction by plutonium dust, as proposed by Tamplin and Cochran and by Geesaman,[10] is correct, then the cancer probability could be 100% for anyone attempting to live in the 150,000-square mile (or greater) fallout zone of the conceived LMFBR explosion. Furthermore, the plutonium dust, because of its extremely long half-life, would forever be in the environment. Presumably, it would be mixed with ordinary dust, kicked up by wind erosion and farming, and blown about and spread by the winds to present a continuous and permanent lung-cancer and other health hazard for any inhabitants of the contaminated and adjacent land.

It is clear, therefore, that the question of the power excursion potential of the LMFBR is extremely serious. We shall now examine the state of the science of predicting the LMFBR explosion potential.

The Basic Theory of Nuclear Explosions in LMFBRs

The LMFBR reactor and coolant system closely resemble the PWR (see fig. 1), except that the fuel rod bundles are each contained in a coolant channel or "duct" as in a BWR (see Fig. 3).

Also, the batch of fuel rods which comprise the core is surrounded by a thick outer ring of rods containing uranium 238 for breeding plutonium; and of each core rod, only the middle vertical section actually contains the fuel, with the top and bottom sections containing U 238. Hence, the core is completely surrounded by a "blanket" of U 238 rods The coolant ducts containing each bundle of rods run the full length of the rods. (See fig. 12.)

The reactor physics and the power excursion theory for an LMBFR are similar to those for a water-cooled reactor; except that liquid metal (heated sodium) is used as a coolant instead of water. The absence of water means that the energetic "fast" neutrons emitted by the atomic fissioning process are not slowed down within the core (it turns out that the fast neutrons enable the breeding process to work); but since fast neutrons are less effective in causing fissioning (see p. 13), the concentration or "enrichment" of fissionable material in the fuel (see p. 12) must be higher in an LMFBR to achieve a critical reactor—over five times higher than a PWR or BWR. This higher fuel enrichment means that should the LMFBR fuel rods be compacted, either by a meltdown or by an explosion which compresses part of the core, the reactivity might not decrease, as it would in a water-cooled reactor, but could *increase*.[11]

The reason for this reverse reactivity effect is that in a water reactor the fuel material, being less enriched in fissionable material, cannot sustain a fission chain reaction (criticality), even if fully compacted, unless the water is present between the fuel rods to slow down the neutrons and thereby increase their effectiveness for causing fission. Fuel compaction in a water-cooled reactor would squeeze out water and thus would reduce the reactivity; that is, shut down the fission chain reaction. On the other hand, the compaction of concentrated LMFBR fuel will *increase* the reactivity, since the nature of highly enriched fissionable material is such that it can be made critical if enough of it is brought together, which the compaction process accomplishes. (An atomic bomb is detonated by compacting extremely rapidly a mass of highly enriched fissionable material with a surrounding TNT-like charge.)

What makes this compaction problem especially serious is that only about 2% core volume reduction, such as could easily occur upon core melting, would raise the reactivity to above the delayed-neutron fraction* and thus trigger a power excursion.[12] Yet, the potential for fuel compaction in an LMFBR is large, since

* Which is about .3% for plutonium-fueled LMFBRs.

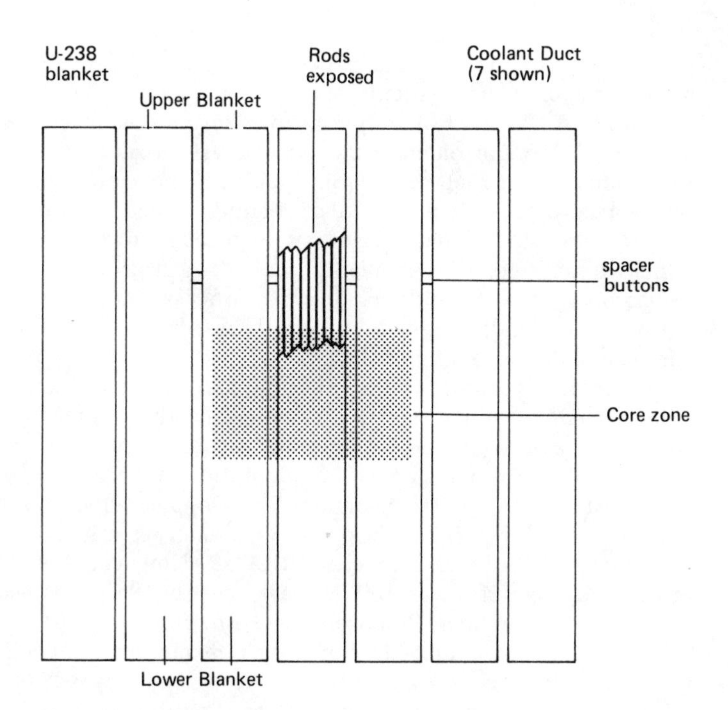

Figure 12. LMFBR core and blanket

only about 50% of the core volume is taken up by fuel rods, and the rest by the coolant. Hence, the core compaction potential is over 50%,[13] should the coolant be expelled or drained, leaving a void for fuel to enter; again, fuel compaction occurs whenever fuel fills voided coolant space between the fuel rods. Also, upon melting, the fuel rods of the core would lose their rigidity, and the fuel would then be easily compressible, as by gravity compaction.

Furthermore, it turns out that the rapid expulsion of the liquid metal coolant, as in boiling, can increase the reactivity in an LMFBR as well, due to complicated nuclear effects. This is in contrast with water-cooled reactors, where a loss of coolant will at least reduce the reactivity and thereby shut down the fissioning. Finally, due to its size, a large LMFBR core will contain several "critical mass" loads of fuel, if fully compacted, so that an explosion due to an initial power excursion might rapidly compact a region of the core enough to make it prompt critical, thereby setting off a secondary, more severe power excursion. In general, the more rapid the core compaction, the greater the rate of reactivity rise and the resultant power excursion.

In short, the LMFBR is prone to *autocatalytic* reactivity accidents—that is, the reactor is its own catalyst for generating power excursions, since fuel overheating due to some malfunction can cause a core meltdown or coolant boiling, which in turn could raise the reactivity and trigger an explosive power excursion, which in turn could conceivably generate disastrous secondary excursions by some rapid recompaction process before the fissioning would be finally stopped by "core disassembly"—blowing the core completely apart by explosion.

Incidentally, the LMFBRs, like the water-cooled reactors are being designed with a negative Doppler reactivity effect (see p. 16), which can safely terminate minor power excursions caused by slight reactivity rises above the prompt critical reactivity level, the threshold for power excursion. The Doppler effect has been demonstrated for LMFBRs in power excursion experiments using a small LMFBR-like reactor called SEFOR. However, the AEC's characterization of these SEFOR tests can be very misleading. Said the AEC in their *Proposed Final Environmental Statement* for the LMFBR program: "In some of the experiments in SEFOR, the reactivity was intentionally increased well beyond prompt critical, and the rapid transient that resulted was controlled by the negative Doppler reactivity effect." [14] This statement can be taken to imply that SEFOR proved that the LMFBR can tolerate strong reactivity rises—"well beyond prompt critical." But this is not true, for the reactivity in the SEFOR tests was *barely* raised beyond prompt critical. Numerically, it was only .06% (% reactivity units) beyond prompt critical,[15] compared to, say, 1% for a severe power excursion.[16] Moreover, the Doppler strength in SEFOR was made three to four times greater than it would be in an LMFBR accident situation.[17] The result was that the fuel temperature in the SEFOR tests rose only about 10%—barely detectable—so that no fuel damage occurred. Hence, the tests did not verify the Doppler effect in the fuel temperature range of severe excursions, which is important to remember.[18] Later, estimates of nuclear explosion accidents in LMFBRs will be given. These estimates will include the mitigating Doppler effect.

Design Basis Accidents

The AEC has argued that the explosion potential is limited and containable, but again, only for certain design basis accidents (DBAs), and not for worse possible accidents (WPAs), namely, worse possible reactor malfunction situations.[19] The DBAs for the

LMFBR[20] include such accidents as the loss of coolant pumping, followed by a failure to SCRAM; or the slow, continuous withdrawal of control rods caused by the malfunction of the control systems, plus SCRAM failure, which increases the reactivity to generate a power excursion. In both of these DBAs, the core overheats. Because of autocatalytic reactivity processes, a severe nuclear runaway (power excursion) can theoretically occur which would heat up and melt the core—which in turn could generate a sodium vapor explosion when interacting with the sodium coolant—or would cause a fuel vapor explosion, if the excursion is stronger. The ultimate effects of such explosion, including conceivable secondary power excursions, must be shown to be containable.

Incidentally, there are two independent SCRAM systems planned for the LMFBRs, so both must fail to function for the above DBA examples. The EBR-1 LMFBR suffered a core meltdown due to an autocatalytic overpower transient when the SCRAM system failed to shut down the reactor even though it functioned. A back-up SCRAM averted a nuclear runaway.[21] In other instances SCRAM systems have actually failed. These incidents, which are described in app. 2, underscore the possibility of SCRAM failure or ineffectiveness.

As with the water-reactor safety analyses, the LMFBR predictions of the DBAs are made using *unverified theory* and include many assumptions that ensure favorable calculational results. For example, let us examine the loss-of-coolant-pumping-type DBA. This DBA assumes that the reactor coolant recirculation pumps stop accidentally while the reactor is at full power, due to some malfunction(s) which cut off the electric power to the pumps. It is further assumed that a SCRAM fails to occur, so that the fission power is not stopped. The core would then continue to generate full power. However, the reduced coolant flow causes the coolant in the core to heat up and boil out of the core, which would leave the fuel to heat up (see fig. 13). The coolant expulsion, or fuel compaction by melting, would then raise the reactivity and generate an explosive power excursion. The AEC's "very conservative" estimate of the explosion resulting from this particular accident is 1,100 pounds of TNT (equivalent), due to a sodium coolant vapor explosion that is assumed to occur when the molten fuel, which is created by the power excursion, eventually interacts with coolant hovering above the core.[22] (Recall that the maximum economical containment capability is about 1,000 pounds TNT.)

However, the AEC has neglected to estimate *secondary excursions* that conceivably could be caused by the sodium coolant vapor explosion. This possibility has been suggested by Argonne National Laboratory, the center for LMFBR safety research, in a report on LMFBR safety problems.[23] It would occur as follows: The power excursion of the above-described DBA melts the core and produces a relatively mild *fuel vapor explosion* within the core (see fig. 13). Molten fuel from the core would then be blown (thrown up) into the coolant chamber above the core. The leading edge of the molten fuel entering the chamber could, by mixing with coolant, generate a sodium vapor explosion. The force of the explosion—say, about 10 pounds TNT-equivalent, which is 1% of the AEC's conservative estimate—could then drive a mass of fuel back down into the core at a high speed, rapidly raising the reactivity by fuel reassembly to generate a disastrous secondary power excursion. According to this author's crude theoretical calculation, this chain of events could conceivably produce a nuclear explosion (fuel vapor explosion) of the order of 20,000 pounds TNT-equivalent,[24] or twenty World War II "blockbuster" bombs (1,000 pounds TNT) combined, which exceeds by far the maximum economical explosion containment capability. Such an excursion would vaporize the plutonium fuel and blow, say, half of it into the atmosphere, as previously discussed.

It is clear, therefore, that computer simulations of LMFBR accidents cannot be arbitrarily stopped after calculating initial power excursions, but must be continued further to calculate whether the core could return to criticality and suffer secondary power excursions before the core explodes completely apart or permanently freezes solid in a shutdown fuel configuration. Recriticality is a problem even after the fuel freezes, because the *afterheat* is much more intense in an LMFBR, due to the tenfold higher core power density, than in a water-cooled reactor. The afterheat could easily remelt the fuel and cause it to slump or fall together into another critical mass configuration. Overall, the industry's theoretical studies of LMFBR accidents have not included a rigorous analysis of all such conceivable autocatalytic reactivity processes.[25]

The recent *Preliminary Safety Analysis Report* (PSAR) for the proposed LMFBR Demonstration Plant* contends that the most likely course of a design basis accident would be for the core, when it overheats and melts partially, to froth (expand), with some fuel being swept out of the core by the flowing coolant. It is contended that these processes will reduce the reactivity enough to shut down the fission reaction (heat) and allow the core

* Clinch River Breeder Reactor PSAR, app. F.

to be safely cooled—that is, to freeze in a stable configuration without exploding, avoiding further trouble. (However, the PSAR neglects DBAs occurring in a *new* core, in which the primary froth mechanism, namely, fission product gases, will not be present.) The difficulty with such speculation is that one can make other plausible assumptions that predict autocatalytic reactivity effects and strong power excursions. The PSAR explores these more pessimistic assumptions for the DBAs, such as fuel rods splitting open near their midheight, which induces the molten fuel within the fuel rods to move initially toward the center of the core, as the fuel seeks to escape the rod through the split. This has the effect of core compaction, which raises the reactivity autocatalytically and produces a fuel vapor explosion.

There are two basic problems in predicting the LMFBR power excursion potential. One involves the extremely complicated fuel and coolant motion within the core under extreme heat and pressure conditions, which vary throughout the core. Complicating the problem are the several sources of pressure (for example, fuel vapor and fuel vapor explosions, coolant flow forces, coolant vapor and coolant vapor explosions, fission product gases, and gravity) and the strong nonuniformities in the composition of the core materials, as well as the difficulty in predicting how and where the fuel will initially break up in an accident—all of these things affect fuel motion. The other problem is in predicting the changes in reactivity and the spatial variations of the core power density resulting from the motion of the core materials. All of these effects and processes are interrelated. Because of this extreme complexity, predicting severe LMFBR power excursions cannot be done, except by making numerous simplifying mathematical assumptions and approximations, the error of which can only be determined by full-scale accident tests—that is, full-scale, integral core-destruct tests, including all reactor systems and components, since they all can affect the motion of fuel and coolant. However, an adequate test program is probably impractical, due to the prohibitive costs and the extreme hazards of such tests, as there would be required a great number of full-scale reactor destructive tests to cover the myriad of different accident initial conditions, which would dictate the fuel motion.

Worse Possible Accidents

Then there are the worse possible accidents (WPAs) that have not been theoretically analyzed for their explosion potential, such as a loss-of-coolant accident (LOCA) in an LMFBR,[26] as in a coolant

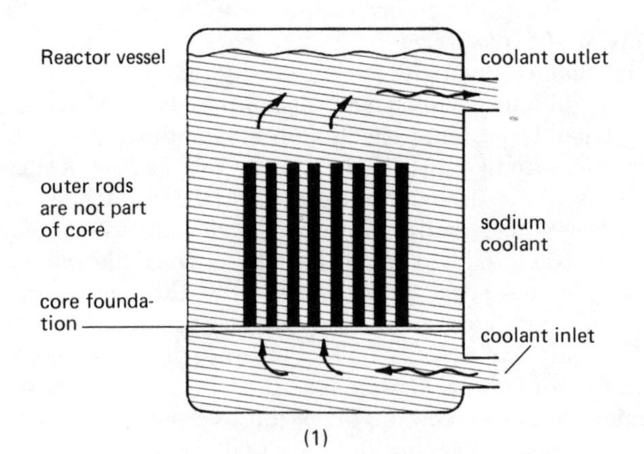

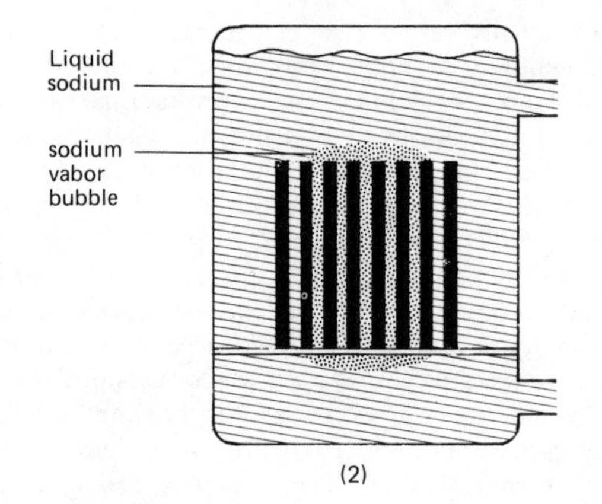

Figure 13.

Sequence:

(1) Reactor at start of accident (coolant flow decreases);
(2) core overheats; sodium coolant boils out of core;
(3) core melts, slumps, and compacts, raising reactivity;
(4) mild power excursion and nuclear (fuel vapor) explosion (this could have been caused by the autocatalytic reactivity effect of sodium expulsion in (2), depending on circumstances);
(5) part of core thrown upward (molten), leading edge interacts with sodium;
(6a) sodium vapor explosion occurs, which propels part of the core downward to rejoin the lower part of the core, causing a rapid fuel reassembly and recriticality;
(7) recriticality, secondary power excursion, about 10,000 lb. TNT-equiv. nuclear explosion.
(6b) Staggered fuel reentry, slow gravity fall.

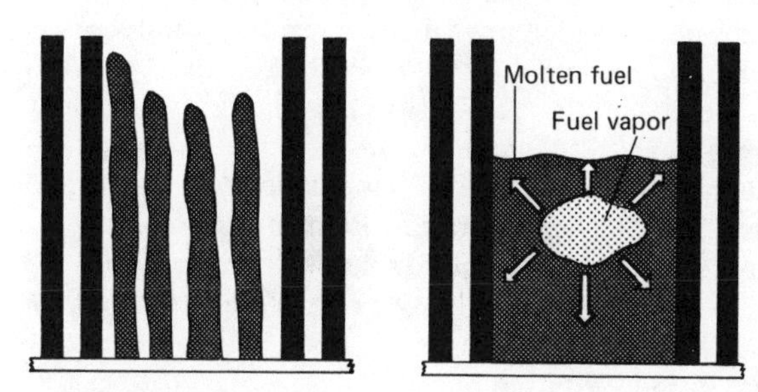
Molten fuel
Fuel vapor

(3)

(4)

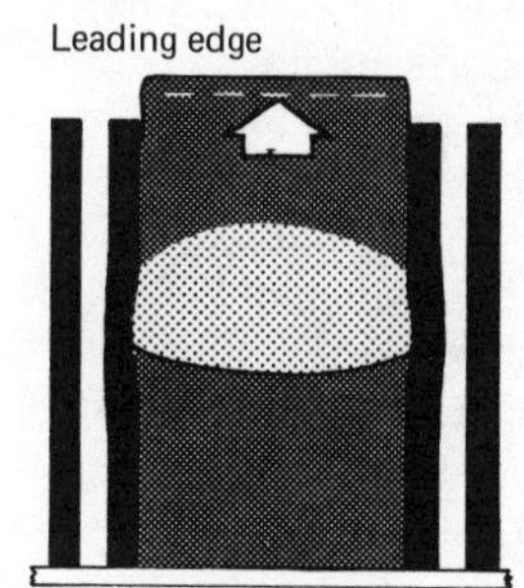
Leading edge

sodium vapor
explosion

(5)

(6a)

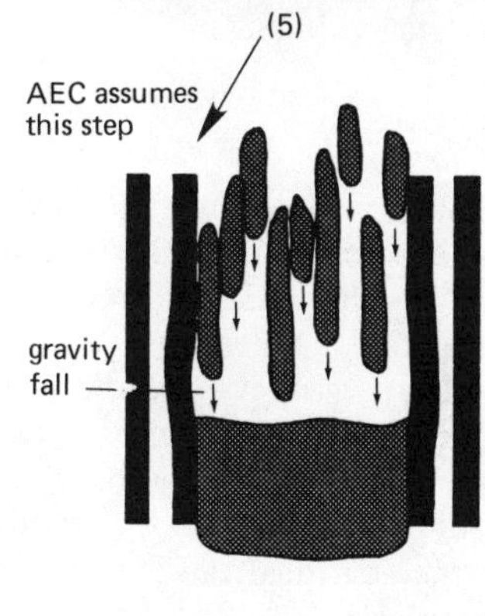
AEC assumes
this step
gravity
fall

(6b)

(7)

pipe rupture, which will produce a more rapid initial core heat-up than the DBAs and, therefore, presumably a more severe initial power excursion. Design measures such as double walls in the vessel and piping could be taken to reduce the probability of a LOCA, as noted in the AEC's environmental statement for the LMFBR program.[27] But again, whether such measures would be "safe enough" is a subjective judgment. Indeed, the proposed design of the LMFBR Demonstration Plant does not include double-walled piping, though the reactor vessel will be double walled.[28] At least the coolant is not highly pressurized as it is in a water-cooled reactor, since sodium has a high boiling temperature, far above normal operating temperatures.

Another WPA could result from the restarting of coolant pumps that have stopped accidentally. The initial loss of coolant flow would cause core melting as discussed above, and the onrush of the returning coolant flow could then rapidly compress the core, made pliable by melting, to generate a severe excursion, which this author estimates at about 6,000 pounds TNT-equivalent or greater.[29] To experimentally investigate the WPAs would require even more full-scale destructive experiments. It is noteworthy that General Electric Co. scientists have stated: "It is, in our view, unlikely that one will be able to design for the worse accident permitted by the laws of nature and end up with an economically interesting system, even after extensive additional research and development has been carried out." [30]

Secondary Power Excursions

The example shown in fig. 13 of a conceivable autocatalytic scenario in the coolant-pump-stop DBA (core compaction by sodium vapor explosion) has been criticized by the AEC as too hypothetical.[31] It is claimed that such explosion scenarios are unrealistic and "of academic interest only." [32] However, there are a number of shortcomings in the LMFBR accident theories used by the AEC to predict a limited explosion hazard, and in any event, the processes are so complex that no theory can be credible without verification by full-scale reactor destruct tests. (No LMFBR reactor destructive tests are planned of any scale.) Because of the fact that the maximum possible explosion has not, therefore, been scientifically established, the maximum conceivable explosion estimates of this author can at least define the seriousness of the matter.

Indeed, an internal document of the Argonne National Laboratory (ANL), prepared by J. F. Jackson and J. E. Boudreau, entitled "Postburst Analysis," reports the results of a much more sophisticated calculation that provides for the sodium vapor explosion mode of fuel compaction and secondary excursion.[33] (The term *postburst* refers to fuel motion following the occurrence of event no. 4 of fig. 13, the initial fuel vapor burst, which is analyzed to investigate the potential for recompaction of fuel by sodium vapor explosion—events 5, 6*a*, and 7.) This ANL calculation, which treats the coolant-pump-stoppage accident, agrees quite well with this author's crude calculation. Specifically, ANL calculated a nuclear explosion greater than 1,160 pounds TNT for this accident in the Fast Flux Test Facility (FFTF) reactor.[34] The containment for the 400-Mwt FFTF is designed to withstand no more than a 150–300-pound TNT-equivalent explosion.[35] Moreover, the internal ANL report did not explore the worst case preditable by the theory used. Scaling the 1,160-pound-TNT figure up for a large, 1,000-Mw (2,500 Mwt) * LMFBR, the theory might calculate a 10,000-pound TNT-equivalent explosion for the coolant-pump-stoppage accident, which supports the author's crude calculation. (The scaling depends on the core mass, which is nine times greater for a 1,000 Mw LMFBR.[36])

With the issuance of the AEC's *Proposed Final Environmental Statement* (PFES) for the LMFBR Program, the AEC has recognized the uncertainty with respect to autocatalytic reactivity effects[37] and reports that theoretical studies are now underway to investigate them. Still, the AEC in their PFES seems to express the belief that uncontainable autocatalytic explosions will have a negligible probability of occurring in any "core disassembly" event under the DBA category,[38] which, incidentally, prejudges the outcome of the planned studies. In support of this belief, the AEC has cited only one reference—a published technical paper, again by Boudreau and Jackson, on "Recriticality Considerations in LMFBR Accidents," which discusses some of the conceivable autocatalytic reactivity effects that could produce secondary power excursions.[39] The AEC quoted a conclusion from this paper: "The results also indicate that a number of effects are expected to be present in realistic situations and can strongly mitigate the reactivity insertion rates." [40] (Lesser rates of reactivity insertion—that is, reactivity increases—would mean a lesser resultant power excursion.) However, this Boudreau-Jackson paper does *not* support the use made of it by the AEC.

Boudreau and Jackson considered in their published paper

* 2,500 megawatts of thermal energy (Mwt) is converted to 1,000 Mw of electrical energy.

what could follow an initial power excursion in which molten fuel is thrown up out of the core. When the fuel falls back down into the core by gravity, assuming no sodium vapor explosion occurs, it would cause a secondary power excursion, since fuel would still be coming together, that is, "reassembling" itself. The speed and quantity of the returning fuel determine the rate of reactivity rise of the reassembly or recriticality process and hence the severity of the secondary excursion. For the reassembly process, Boudreau and Jackson *hypothesized,* based on some argumentation, that the returning fuel would be "staggered," that is, the fuel could fall by gravity in a train of small pieces rather than one large mass, which would produce a more gradual reactivity rise and therefore a lesser secondary excursion. (The technical name for this staggered fuel motion is "noncoherent fuel reentry"; see event 6*b* in fig. 13.) This is, essentially, the "realistic" result on which the AEC relies.

Even if we accept this *staggered fuel reentry* hypothesis, the AEC neglects the fact that the secondary power excursion would still be worse than the initial one; so the accident isn't over and must still be followed to determine the ultimate course of the fuel motion. For example, the second excursion would cause the core to explode (fuel vapor explosion), which would compact parts of the core as they slam into surrounding structures inside the reactor vessel. This recompaction may cause a more severe, *third* excursion, which might be disastrous.

In regard to the staggered fuel hypothesis, who is to say whether this is any more realistic than a more massive, coherent fuel reentry, as depicted in event no. 7 of fig. 13, since we have no experiments of whole core destructive tests to tell us what is *real?* For the case of *coherent* fuel reentry by gravity only, Boudreau and Jackson calculate disastrously high rates of reactivity increase, which, incidentlly, are consistent with this author's past calculation.[41] In any event, even if we performed such experiments, we couldn't detect the detailed fuel motion to verify our theory—another problem, to be discussed later. In fact, Boudreau and Jackson did not draw any definite conclusions in regard to the fuel motion, contrary to the AEC's implication. They pointed out repeatedly throughout their paper the many sources of large uncertainty and the extreme difficulty in attempting to calculationally predict the fuel motion, and they concluded that coherent motion (large reactivity insertion rates) cannot be ruled out. They emphasized: "We are not now drawing any definitive conclusions about the behavior of a real reactor." [42]

These analysts have, however, offered their "engineering

judgment" that for the case of gravity fall, staggered fuel reentry is "likely" for *small* LMFBRs. Accordingly, they express the judgment that the rate of reactivity increase during the fuel reentry will be much less than that calculated for coherent fuel reentry (gravity fall), although the rate would still be about ten times that of the initial power excursion, which could make the secondary excursion worse than the first, as mentioned earlier. But they admit that their judgment is "subjective," which, of course, is not reliable for safety purposes. (They added that they "have not made such judgments for larger LMFBRs to date." [43])

The limited reactivity rise rates calculated by Boudreau and Jackson in their published paper due to incoherent fuel motion apply only to *gravity fall* of the fuel. They briefly noted, however, that if the fuel is *shot* back into the core by a sodium coolant vapor explosion, as is our main concern here, then high rates of reactivity increase can occur, even if the fuel reentry is staggered.[44] (The values they report would produce a disastrous 10,000-pound TNT-equivalent *fuel vapor* explosion for a large LMFBR, as before estimated.) They did not elaborate further, nor did they reference their internal ANL report, "Postburst Analysis," which contains their analysis of the sodium vapor explosion mode of core compaction.

The AEC did not address the possibility of autocatalytic core compaction by a sodium vapor explosion in their *Proposed Final Environment Statement* for the LMFBR. However, in an unpublished commentary on the internal ANL report of Jackson and Boudreau, the AEC asserted that sodium vapor explosions are "considered very unlikely," given molten fuel contacting sodium coolant, and that, therefore, this form of autocatalysis "is not considered to be credible." [45] For their basis, the AEC referenced the work of H. Fauske, who has proposed a "superheat theory" of coolant vapor explosions which predicts that the mixing of molten UO_2 fuel with liquid sodium will not produce a sodium vapor explosion under LMFBR reactor conditions.[46] (Presumably, this applies to plutonium oxide fuel as well.) It thus remains to examine the possibility for sodium vapor explosions in LMFBR fuel melting accident situations, in order to assess the credibility of severe autocatalytic fuel compaction by sodium vapor explosion.

The limited power excursions deemed "realistic" by the AEC for the DBAs, which neglect the various possibilities for severe autocatalytic reactivity effects, are not predicted to be severe enough to generate significant amounts of fuel vapor with respect to the amounts and pressures required to cause a fuel vapor explosion that could threaten containment rupture. Nevertheless, the heat of the excursion is predicted to make the core substantially molten. The concern is that, when the molten fuel then mixes with the liquid sodium coolant, the sodium will heat up and boil explosively; that is, a sodium vapor explosion will be generated. The containment systems of LMFBR designs are usually designed to withstand such sodium vapor explosions, although this mode of explosion is calculated conservatively to approach the limit of the economical containment capability for the DBAs.[47]

Although the AEC, in the PFES, reasserts the "superheat theory" to practically rule out the possibility of sodium vapor explosions,[48] it has not been *completely* ruled out. Accordingly, tentative design provisions are being made in the containment of the LMFBR demonstration plant for withstanding sodium vapor explosions that might result from molten-fuel-coolant interactions during a DBA.*, [49] To be consistent, however, one would also have to determine the potential for autocatalytic power excursion by core recompaction following the assumed sodium vapor explosion.

The possibility of sodium vapor explosions is not so remote as suggested by the AEC. Sodium vapor explosions have actually been observed in small laboratory tests in which molten uranium oxide fuel was interacted with liquid sodium.[50] This experimental fact requires that the matter be examined in more detail. The following analysis will show that large-scale, integral core destruct tests would be needed to settle this question.

The basic materials of an LMFBR are the fuel, structural steel (for example, fuel cladding), and sodium coolant. In a power excursion the fuel can be heated and melted, which in turn can heat up and melt the steel (steel melts at a much lower temperature than the fuel). The "superheat theory" predicts that sodium vapor explosions are not possible in the case of molten uranium oxide

* The PSAR for the proposed LMFBR Demonstration Plant now considers some limited forms of autocatalytic reactivity effects—but not fuel compaction due to a sodium vapor explosion—and calculates that fuel vapor explosions could possibly occur which would threaten containment rupture (app. F). As for the possibility of sodium vapor explosions compounding the explosion, the PSAR now assumes that they won't occur, which is a matter yet to be resolved. Incidentally, should the containment be faulty and fail to contain such limited DBA fuel vapor explosions, about 8% of the plutonium fuel (200 lb.) could escape into the environment, to say nothing of the fission products. (See A. B. Reynolds et al., "Fuel Vapor Generation in LMFBR Core Disruptive Accidents, *Nuclear Technology* 26:165, 170.) Such plutonium release could require permanent abandonment of 10,000 square miles of land—a figure arrived at by reducing previous estimates proportionally.

fuel interacting with sodium coolant, when the sodium contains impurities, but are possible in the case of *molten steel* interacting with sodium, even with impurities in the sodium. This superheat theory requires that the sodium be heated above its boiling-point temperature without actually boiling—that is, "superheated"—only to flash instantly (explosively) into vapor when sufficiently agitated. According to the theory, impurities, such as solid particles, would promote gradual nonexplosive boiling and thus avoid superheating the sodium in the case of molten fuel. (The fact that sodium vapor explosions were actually observed in small laboratory tests is attributed to the lack of impurities in the test materials.) Steel, however, being metal, gives up its heat more rapidly than oxide fuel, which is a ceramic material like Corningware. Because of this, sodium can theoretically be superheated by molten steel and explode, regardless of impurities, provided the molten steel is hot enough.[51]

The significance of molten-steel-induced sodium vapor explosions is that these explosions could be the catalyst for unleashing the sodium vapor explosion potential of the molten fuel. Initial steel-induced sodium explosions could pulverize the nearby molten fuel into ultrafine droplets which would then spray (disperse) into the sodium and boil it so rapidly by intimate mixing that a larger, full sodium explosion would be produced by the molten fuel without the need for superheating the sodium. This would make the "superheat theory" irrelevant to molten fuel. Such a catalytic preexplosion, which triggers the main explosion, is believed to be the way in which coolant explosions occur.[52] Indeed, as we shall see shortly, the molten steel by itself, not as a catalyst for the molten fuel, could generate a strong sodium vapor explosion.

However, the AEC argues that, even with steel, sodium explosions are unlikely.[53] They argue that a layer of steel vapors (steel turned to vapor by the intense heat) will form on the molten steel surfaces and will be sufficiently dense at the high, core-melting temperatures to insulate the steel from the sodium, when the two materials interact, thereby preventing rapid, explosive boiling of the sodium by the molten steel. For high enough steel temperatures, this argument seems plausible, though it is still only a hypothesis. However, it also appears possible that the temperature of the molten steel can be low enough to avoid a dense, insulating layer of steel vapor and yet may still be high enough to produce a sodium vapor explosion, by the superheat theory.[54] Thus, there is uncertainty as to the minimum-threshold steel temperature required by the superheat theory to cause a sodium explosion. Said H. Fauske, who is the chief exponent of the superheat theory in the LMFBR safety field, the threshold condition of sodium explosions for steel "remains largely unknown." [55] And so it appears that the theoretical possibility for sodium vapor explosions by molten steel depends on whether the specific temperature values of core materials will be within a critical range (not too high and not too low), which must be quantified to assess the likelihood of explosions under the theory. Unfortunately, the AEC does not quantify the matter, nor does it document its case. Furthermore, Fauske does not assert the AEC's view; he has not ruled out the possibility of sodium vapor explosions caused by molten steel.[56] Indeed, he concluded that steel-sodium interactions are "potentially much more explosive" than fuel-sodium interactions.[57]

The possibility for catalytic, steel-induced, sodium explosions will depend, therefore, on both the steel and sodium temperatures being within required, but unknown, ranges. However, these temperatures will vary throughout the mix of core materials, depending on the way they mix with the hot molten fuel and on the time spent in mixing, as each material conducts and holds heat differently; and all of this will depend on the complex motion of the various core materials. The net result of such complex interactions following a power excursion are too difficult to predict with any confidence. Because of this, and because the sodium explosion threshold condition is largely unknown for molten steel, there is no way to reliably predict the occurrence of sodium vapor explosions or to prove their impossibility under the superheat theory, other than by integral core destructive experiments. Consequently, one would have to resort to such experiments, which may have to be full-scale to ensure that all conditions which could affect the temperatures of the various materials are present. Furthermore, because of the element of chance involved due to complex interrelationships among initial conditions and other variables, at least several core destruct tests for a given accident would have to be performed if sodium explosions are not observed at first, assuming one could differentiate between fuel vapor and sodium vapor explosions, should an explosion occur. These tests would have to be further multiplied to cover the various possibilities for reactor system malfunction, namely, the various DBAs and, if desired, some or the worst of the WPAs. Of course, such experiments to investigate the possibility of sodium vapor explosions would investigate other matters at the same time, such as the various conceivable autocatalytic reactivity effects, including, of course, fuel compaction by sodium vapor explosion if sodium explosions occur.

The fact is that sodium vapor explosions have been produced with molten UO_2 without the need for the molten steel catalyst in small-scale, nonreactor experiments.[58] The superheat theory allows for this fact, however, provided the materials do not contain impurities, as mentioned before, which was the case for the nonreactor experiments.[59] The AEC points to more realistic experiments in which one or a few (seven) LMFBR fuel rods with steel cladding were excursion heated in a tube of sodium placed in the core of the TREAT test reactor.[60] No sodium explosions were observed, due assumably to impurities and/or differences in the manner of contact among the materials. However, these experiments cannot be reliably extrapolated to whole-core accident situations of 100,000 closely packed fuel rods in which molten core materials are generated.[61] Tests with only a few fuel rods might not trap the sodium sufficiently to heat it up to explosive pressures, or the steel cladding may not have had time to heat up enough, since it would quickly contact cold surfaces of the small test chamber. (Incidentally, when such negative experiments are reported, they are never adequately analyzed by the laboratories for their implications or validity.[62]) Moreover, the TREAT tests apply only to the type of accident in which the power excursion occurs when the fuel rods are immersed in sodium coolant. We, however, are presently concerned with the situation studied by Jackson and Boudreau in "Postburst Analysis": the coolant-pump-stoppage-type DBA in which most of the sodium is boiled out of the core *prior* to the excursion (see fig. 13). The molten fuel-sodium interaction subsequently occurs in a manner wholly different from the TREAT tests.

Sodium vapor explosions, therefore, if they can occur in an LMFBR accident under the superheat theory through the steel catalyst, will depend on the amounts, composition, configuration, temperatures, and pressures of the materials and surroundings, which cannot be reproduced except by integral core destruct tests. The question is, then, whether *small-scale* integral tests would suffice (even these are not planned, however). Small-scale integral tests would be illuminating but would not be conclusive, should explosions not occur, for it may take large-scale amounts of core materials to achieve the proper heating conditions. Considering the matter still further, it would seem that the integral core tests would have to be *full-scale,* in order to duplicate the geometric configuraion of the reactors, which dictates the crucially important motion and heating of the various core materials. Moreover, the potential for sodium vapor explosion will depend on the degree of the *fuel heating,* which depends on the energy yield of the preceding power excursion(s). To determine the energy yield for each accident under investigation would alone require full-scale tests, in order to produce all of the reactivity effects that determine the severity of the power excursion. On the other hand, if small-scale tests exhibit sodium explosions, then full-scale tests would absolutely be needed to investigate the reactivity effects of the fuel motion (possible rapid recompaction) caused by a sodium explosion, due to the nature of *reactivity*, which only full-scale fuel loadings can investigate.

Furthermore, the *superheat theory*, on the basis of which the AEC downplays the possibility for sodium vapor explosions, has not been established. It is just one of many theories of coolant vapor explosion that can explain the coolant explosions that have been experienced.[63] For example, LMFBR safety analysts from Germany concluded that there still "remains a further mechanism which may lead to explosive fuel-sodium interaction." [64] Nor has Fauske drawn any final conclusions on this matter.[65]

Incidentally, no sodium vapor explosion, evidently, had occurred in the fuel meltdown accidents in the EBR-I and Fermi LMFBRs (see app. 2). However, this fact does not rule out the possibility of sodium vapor explosions, for in each case there was no explosive, super-prompt critical power excursion to pulverize the sodium and molten fuel (by a mild fuel vapor explosion) so that the rapid, thorough mixing of these materials that is needed to produce a sodium explosion could be attained.[66] Also, the fuel was metallic in EBR-I and Fermi, which is not the same as the oxide fuel presently planned for use in LMFBRs. Metallic fuel has a much lower melting temperature than oxide fuel, and this may have contributed to the fact that no coolant vapor explosions were observed in these accidents, since lower temperatures considered alone would have a lesser coolant explosion potential.

It can be concluded, therefore, that the sodium vapor explosion-core compaction mode of autocatalytic power excursions, which theoretically can produce disastrous 10,000-pound TNT-equivalent fuel vapor explosions, or greater, cannot be ruled out, that any claim of a very low likelihood of such excursions is mere speculation, and that integral core destruct tests would be needed to settle the question.

Other Modes of Nuclear Runaway

There are other, equally serious, modes of conceivable autocatalytic reactivity effects, which do not require a sodium vapor explosion. These include such processes as:

1. Compaction of fuel regions. In this case, the fuel violently expands by an initial power excursion and then slams into the structure tightly surrounding the core, which results in regional compaction (piling up). Based on Boudreau and Erdman's calculations, such a process might increase the reactivity at a rate of about 1,000$ per second to trigger a 10,000-pound TNT-equivalent secondary excursion.[67] (1$ reactivity defines prompt critical.)

2. Reduction in neutron "streaming." In some types of accidents, such as the DBA considered in detail above, the initial power excursion occurs after most of the coolant has been expelled from the core region of the reactor by boiling. This means that the space between the fuel rods would then be emptied of coolant. When the power excursion occurs and the individual fuel rods explode due to the excursion, the space between the rods becomes filled with expanded fuel, which blocks the stream of neutrons that would otherwise flow up and down these spaces. This in turn reduces neutron leakage from the core and raises reactivity, which would amplify the excursion and might produce a 9,000-pound TNT explosion, according to this author's crude calculations, which need more refined analysis and, of course, experimental verification.[68]

3. Fuel slug rebound or crashdown. In this case, pieces of the core, after first being blown apart, fall or rebound back into the core. Again, 10,000-pound TNT excursions are conceivable.[69]

4. Another conceivable autocatalytic reactivity effect is fuel "implosion" during a power excursion, that is, fuel blowing into a cavity in the core (formed earlier in the accident) in such a way as to move fuel toward the core center or otherwise reduce neutron streaming leakage, which rapidly raises reactivity.[70]

In these processes, the estimated 10,000-pound TNT explosion is due to exploding *fuel vapor*, generated by very severe power excursions—that is, unaided by any sodium vapor explosion. Therefore, proving that sodium vapor explosions are unlikely or impossible would not eliminate the possibility for catastrophic nuclear explosion.

Lack of Theoretical Bounds of Explosion Potential

These autocatalytic possibilities, or at least some of them, are now being studied by computer simulation.[71] As we know, however, the fuel motion problem is extremely complex. Moreover, complicated forms of the neutron flow theory must be used to calculate the all-important *reactivity,* given the fuel motion, namely, transport theory, which makes predicting the course of accidents even more difficult. Furthermore, the fuel motion and reactivity calculations are interrelated, so that errors feed on each other. It is therefore difficult to escape a conclusion that no prediction, not even for the DBAs, could ever be believable. Consider the opinion given by an authority on LMFBR design for the General Electric Co., K. P. Cohen: "[W]e don't know very much about what the meltdown accident is going to be, and though one can indeed make calculations about it, one would be naive to really believe them." [72] Recently a manager in the LMFBR safety research program at the AEC's Argonne National Laboratory admitted that "mathematically derivable upper limits [of the LMFBR explosion potential] are not obtainable." [73] Indeed, the AEC has admitted that "a completely deterministic or mechanistic analysis of the disassembly process is not feasible," [74] which means the same thing.

In place of mathematical bounds or upper limits of the explosion potential of the DBAs, the nuclear community is seeking to calculate something called "defensible bounds," based on *subjective* judgments of the possible courses of fuel motion.[75] These defensible bounds are to include judgments of the probability of the true explosion potential being greater than the "defensible" limit. These judgments are to be formed on the basis of "understandings" and "experience" as to how the fuel will probably move during an accident. These "understandings" are to be developed by repeated calculations in which the different assumptions are varied;[76] but obviously the understanding will be only as good as the assumptions, since they will not have been verified by integral core destructive tests. Therefore, no matter how one views it, the subjective "engineering judgment" will simply be another name for *hypothesis* and therefore unreliable.

Still, it remains to be seen what will be the magnitude of the "defensible" explosion estimates. When these are presented, it will then be necessary to evaluate the calculations and to identify the crucial assumptions. For example, in support of an earlier claim of a low explosion hazard, the AEC stated to Congress that the VENUS computer model for calculating LMFBR explosion accidents "takes into consideration autocatalytic reactivity effects such as fuel motion." [77] However, a close examination of the VENUS theory shows that it assumes a simple reactivity theory which breaks down when the fuel movements become large.[78] The AEC now recognizes this shortcoming.[79]

Reactor Destruct Experiments: Needs and Practicality

Clearly, integral core destructive tests (full-scale) would have to be performed before any theoretical prediction of the explosion hazard could be believed. The full-scale aspect is dictated by (*a*) the various reactivity effects, which can only be produced by full-size fuel loads; (*b*) the fuel motion, which depends on the size, and (*c*) the sodium vapor explosion possibility. This necessity also follows from the Newtonian *experimental philosophy,* which was previously discussed in connection with water-cooled reactor accidents. (It is noted that the military arm of the AEC deemed that the Cannikin H-bomb test for the ABM warhead in Amchitka, Alaska, had to be full-scale; the H-bomb theory undoubtedly has similar needs for full-scale verification.) For the case of water-cooled reactors, there have been at least small-scale integral core destruct tests, though only for metallic fuel, not for the presently used oxide fuel. These were the BORAX and SPERT tests discussed earlier (pp. 67–68). For the LMFBR we have not even had the BORAX or SPERT equivalent of a test. Incidentally, Hans Bethe, Chauncey Starr, and Walter Zinn, who are respected figures in the nuclear community and known advocates of nuclear energy, have concluded in an AEC-funded study that "large scale" testing "similar to SPERT" is "essential" to resolve LMFBR safety issues.[80] (Since SPERT was actually a *small-scale* integral core test, they probably meant "large scale" relative to TREAT-type excursion testing of a few fuel rods and did not mean to suggest large-scale integral tests relative to the full-size, large commercial LMFBRs being planned.) However, the suggested SPERT-type testing—small-scale integral core destruct tests—is not even included in the NRC's safety research plans.[81]

Small-scale integral core destruct tests would be very useful, for they could throw much light on the various conceivable autocatalytic reactivity processes, such as neutron streaming effects and core compaction by sodium vapor explosion, besides providing other data. And, of course, they *may* not be too hazardous. Unfortunately, small-scale tests cannot investigate the conceivable autocatalytic reactivity effects associated with full-size LMFBRs, namely, regional core compaction, since small-size cores cannot duplicate the reactivity effects of large cores.

In a published report, *Safety Problems of the LMFBR,* ANL stated that the question of autocatalytic core compaction "conceivably could benefit from a proof test of a complete whole-core destructive-excursion experiment." [82] However, the report does not quantify the significance of such autocatalysis or the potential for occurring. In a limited-distribution ANL report on "Studies of LMFBR Safety Test Facilities," it was stated: "The need for integral core destruct tests will be examined in greater depth in a later phase of the safety test facilities study program." [83] However, that "facilities study program" has been "terminated." [84] The issue of the necessity for integral core destruct tests is completely neglected in the AEC's *Proposed Final Environmental Statement* for the LMFBR program (PFES).

Recently the NRC's director of reactor safety research, H. Kouts, noted that:

> there is a growing belief that the fast [neutron] reactor program will require large tests of the course and consequences of hypothetical core disruptive accidents.
>
> Such tests are not yet scheduled, but they have been introduced into both ERDA and NRC planning.
>
> In fact, it is likely that at least one and probably several large test facilities will be needed to provide better information on fuel-coolant interactions, the conversion of thermal energy to mechanical work in such an interaction [i.e., sodium vapor explosions], and basic information related to the possibility and consequences of recriticality by recompaction following a disassembly.[85]

To the layman, this may seem to indicate that the NRC is planning adequate experiments, or at least SPERT-type integral core destruct tests. However, Dr. Kouts subsequently stated that he did not mean that *integral core* destruct tests are being planned.[86] Instead, the NRC plans only a somewhat larger version of the present TREAT test reactor. TREAT is a small test reactor which subjects fuel rod specimens to power excursions in order to study the dynamics of the fuel motion right after the bursting of the cladding. Up to seven fuel rods are tested in a special chamber or "capsule" placed inside its core. The newer facility will simply use a larger test chamber to accommodate a larger number of fuel rods, but the amount of fuel in the chamber will not be enough to achieve criticality on its own. Rather, the test fuel will undergo fissioning by neutrons flowing in from the core of the test reactor, which is designed to produce power excursions without damage. Thus, reactivity-fuel motion interrelationships will not be present, except incidentally, since the test fuel will have an incidental effect on the reactivity of the test reactor. Therefore, though such tests would provide some information for developing some of the theory for predicting LMFBR accidents, they would not be adequate for verifying overall theoretical estimates of the explosion potential, because of the reasons given earlier. Also, the extent of the usefulness of these "super-TREAT," nonintegral experiments cannot be determined until they are described in detail in the literature. By planning such tests, however, the NRC effectively admits that the present explosion theories are unreliable. It will be interesting to see whether the NRC approves the construction of the LMFBR demonstration reactor plant, now that it is planning these experiments.

The practicality of full-scale integral core destruct testing is, of course, extremely doubtful, even if confined to the DBA-type accidents. Many such tests would be needed to cover (1) the various initial states of the reactor, which will determine the fuel motion (fuel burn-up, fuel bundle arrangements, initial power level and coolant temperature, and so on, and some combinations thereof); (2) the various accidents and degrees thereof; and (3) the size of the LMFBR investigated, since projections range from the initial 300-megawatt LMFBR demonstration plant to 3,300-megawatt plants for the future. Moreover, the hazards of just one test would be awesome, since tons of poisonous plutonium fuel would be involved, plus the fission products. The radioactive clean-up problem after each test would be formidable as well. All integral tests would have to be conducted well underground, to ensure adequate containment of possible nuclear explosions.*

Large numbers of such tests would be required, since the internal workings of the theory of each accident could not be verified; only the end results could be checked with prediction. This is because one could not measure the detailed fuel motion occurring during the progress of a test, for any detection system placed

* It should be noted that the partial nuclear weapons test-ban treaty prohibits the United States from undertaking to carry out "any nuclear weapon test explosion, *or any other nuclear explosion*... in any... environment if such explosion causes radioactive debris to be present outside the territorial limits" of the United States (D. R. Inglis, *Nuclear Energy: Its Physics and Its Social Challenge* [Addison-Wesley, 1973], p. 357; emphasis added). Fairly interpreted, this treaty would seem to require that, if any LMFBR core destruct tests were conducted, every necessary precaution would have to be taken to ensure that virtually no leakage of radioactivity to the atmosphere occurs, including containment protection against the worst theoretically conceivable explosion that could arise in the test.

in the core would affect the fuel motion and reactivity greatly and make the results meaningless.

Recall the contention of the LMFBR Demonstration Plant PSAR (app. F) that the most likely course of design basis accidents, or "core disruptive accidents," is early fuel frothing and/or expulsion from the core which will reduce the reactivity so as to safely shut down the reactor, like a safety fuse. If there is great confidence behind this contention, then it would be relatively simple to build a full-scale core disruptive accident test reactor (well underground, of course) to verify these expectations. Since no explosions would occur, it would be fairly easy to test varying DBA conditions by using replacement cores, like cartridges. The difficulty with such experimentation, however, is that full-scale LMFBRs would have to be operated for several years to generate test cores having the various possible fuel burn-up conditions, a risk we may not want to take, since the "safety fuse" prediction would not yet have been verified. Then, too, the test reactor itself could suffer a mishap worse than a DBA; or we might find a DBA test disastrously underpredicted. Such are our dilemmas with the LMFBR.

It is still conceivable, though doubtful, that the theory and calculational capabilities could be developed with the aid of small-scale integral core destruct tests such that upper-limit-type analyses will *predict* absolutely that disastrous autocatalytic explosions are not possible for the DBAs. For this to be possible, the small-scale integral tests would at least have to exhibit no sodium vapor explosions; in that event, the requirements for full-scale testing (numbers of tests) could be reduced to where it may be practical, though formidable. (Full-scale tests would still be needed to verify the theory.) This scheme of research will be referred to as the "minimum research program." Society, however, may not want to accept the risks and expense of such a minimum program; the costs are guessed at about five billion dollars.

It should be noted that such a minimum experimental program was proposed by J. B. Nims of the Atomic Power Development Associates (designers of the Fermi LMFBR) at the 1963 Argonne National Laboratory conference on the safety of breeder reactors.[87] Nims proposed a series of partial core meltdown experiments, short of explosion, to learn the manner in which the core would melt down; then, with a more confident understanding of core meltdown, full-scale reactor meltdown tests would be designed and performed to determine the severity of the explosions associated with the previously established core meltdown patterns. The original *LMFBR Program Plan,* which provided for studies of the necessity for explosion testing, adopted the Nims method for consideration;[88] however, the studies have not materialized.

Choices for Society

The necessity for full-scale accident tests to verify accident analyses and the doubtful practicality of such tests confront society with the following choices: (1) abandon the LMFBR to avoid the uncertainty of unverified safety calculations, assuming that containable explosions will still be predicted for the DBAs after the more rigorous theoretical studies into autocatalysis are performed; (2) build the LMFBRs without establishing the explosion potential for any manner of core melting event in the hope that accidents will be prevented or contained in case a DBA occurs; or (3) postpone the LMFBRs in favor of the abovementioned minimum research program to determine as much as practical whether the DBAs can be safely contained in practical LMFBR containment systems.

If society favors the third alternative, there still remains the question of the acceptability of the risks of WPAs. It may very well be that straightforward theory will predict severe power excursions with disastrous *fuel vapor* explosions for these accidents without the need for investigating and including complex autocatalytic reactivity effects in the theory. For example, this author estimates a disastrous 6,000-pound TNT explosion for the coolant pumping restart situation (see p. 128). In this case society will have to judge the likelihood of the WPAs and to decide whether to accept these risks. In considering choice (2), society will have to judge the safety of relying on accident prevention and on the subjective judgments of the accident analysts in the nuclear community that the probabiliy of fuel motion in a DBA leading to an uncontainable explosion is small. Actually, the level of safety of choice (2) may be greatly lowered in the future, in view of T. B. Cochran's cost-benefit analysis of the LMFBR program, which shows that the projected economic feasibility of the program depends on doing away with explosion containment altogether after the first few large LMFBRs, in order to reduce costs.[89] The feasibility predictions are also based on building even larger-size LMFBRs to take advantage of economies of scale.[90] However, larger core sizes aggravate the potential or uncertainty for autocatalytic excursions[91] (and would make correspondingly larger integral core destruct tests necessary). Indeed, the AEC is seeking to eliminate explosion containment for even the LMFBR demonstration plant (the 300-megawatt Clinch River LMFBR) on the basis of judgments that accident preventive measures will make the likelihood of a serious power excursion or core meltdown accident remote.[92]

Accident prevention. In regard to accident prevention, experience shows that LMFBRs are prone to core meltdown and mishaps which could lead to nuclear runaway (see app. 2, nos. 1 and 2). Of course, lessons have been learned, and more attention is being given to the design of the LMFBR and its safety systems to prevent accidents—that is, to minimize the probability of uncontrolled situations. The adequacy of the preventive measures remains to be examined; however, to do so thoroughly would be a formidable task, due to the quantity and complexity of detailed information that would have to be evaluated. The present analysis is offered without an extensive evaluation of the preventive measures, because it is believed that the question of the explosion hazard is so profoundly fundamental, due to the plutonium, that it must be thoroughly studied first. That is, there is a need for society to first make the fundamental decision of whether it wants to consider the LMFBR at all, if there is a physical possibility of a catastrophic nuclear explosion. It was only four years ago that the AEC, when it decided to proceed with the LMFBR demonstration plant, gave assurance to the public in the *Environmental Statement* for that reactor that the provisions in the reactor containment structure for "blast and missile protection within the inner barrier provide substantial margins against major potential energy releases [explosion] for *all* classes of accidents"; and that "[w]hile it is impossible to postulate with precision the detailed course of accidents, including their likelihood and possible environmental consequences, *it is possible to place bounds on such accidents*" [93] However, upon close study we find that these assurances are scientifically unfounded; and they have since been withdrawn.

Since core meltdown and severe power excursions cannot absolutely be prevented, and since an upper limit of the probability of such accidents cannot be definitely determined, the likelihood of serious accidents occurring will ultimately be a matter of personal judgment. Furthermore, there is surely substance to the view: "If something can go wrong, it will." Hence, society may wish to resolve the explosion hazard question first. Of course, just pursuing the question, given the possibilities for reactor system malfunction, would be a formidable research undertaking, as explained earlier.

In light of the enormous potential benefits of the LMFBR as an energy source, it may very well be that society will want to weigh the explosion hazards and uncertainties relative to the accident preventive measures. One can rest assured that great care is being given to designing and developing an impressive, but very complex, set of preventive measures, including early warning systems to detect trouble and then to actuate alarms and safety shutdown systems (SCRAM) automatically, well before the reactor can get out of control.[94] For example, if the coolant flow to one of 200 fuel rod bundle ducts in the core becomes blocked by a foreign object, then the affected fuel might melt down and eventually cause the reactivity to rise. Without a SCRAM, the power could then rise, causing the whole core to overheat and bringing on a major accident.[95] To guard against this, design measures are being taken to minimize the likelihood of flow blockage and to monitor the core behavior to detect a flow blockage event. The monitoring instruments would include (1) devices to detect radioactivity leakage from the affected fuel, excessive coolant temperatures, reduced coolant flow, and pressure disturbances; and (2) computer systems to process the data of these instruments and interpret what may be happening inside the reactor and to actuate the SCRAM when necessary.[96] (Incidentally, a flow blockage event without SCRAM is not included among the DBAs; it remains to be analyzed.)

Also, the LMFBR Demonstration Plant reactor will have two separate SCRAM systems to respond quickly to sudden breakdowns that could lead immediately into a major accident.[97] Recall that a SCRAM is the main protection in the LMFBR against serious accident; more specifically, in order for a DBA and many WPAs to occur, a failure to SCRAM must occur following the initiating malfunction.

The question of the reliability of these preventive measures, of course, would have to be thoroughly treated before any affirmative decision on the LMFBR could be made. For the time being, a few critical remarks on this subject would be appropriate.

The two-SCRAM protection may not be as reassuring as it might appear, because of the autocatalytic potential of the LMFBR, which means that relatively minor, high-probability malfunctions could lead into disastrous situations should a SCRAM fail to occur. That is, the DBAs described earlier (pp. 123–24) are not uniquely serious but are only representative of many relatively likely accident possibilities. Thus, to consider accident likelihood, one would need to appreciate all malfunction possibilities with respect to the need for a SCRAM. In regard to SCRAM reliability, it should be noted that in past instances a SCRAM system has failed, another has been found inoperative, and another had been ineffective even though it functioned; and operators have failed to take appropriate SCRAM action at other times (see app. 2, nos. 1, 2, 4, and 5). Also, though the back-up SCRAM system for the LMFBR Demonstration Plant will be different in design than the primary system, in order to minimize the likelihood of common defects, the FFTF reactor apparently will not,[98] which adds to the risk of that reactor.

The primary purpose of the LMFBR Demonstration Plant relative to safety is "to obtain the necessary operating experience" to test out the feasibility of the complex accident preventive measures.[99] But what if these measures fail, or unforeseen phenomena occur, and a core explosion results? The demonstration reactor will have about 1.3 tons of plutonium[100] which, if half were released to the atmosphere, could conceivably require abandonment of a land area of the size of Virginia. So the explosion hazard question will have to be resolved even before the demonstration reactor is built. Actually, the FFTF, which is about to be operated, is in many respects prototypical of the LMFBR Demonstration Plant reactor and hence its operation would be equally uncertain as well. With its one-half ton of plutonium, the FFTF poses a similarly serious contamination hazard for Montana, the Dakotas, and Minnesota, and possibly other adjacent areas.

Past experience in operating LMFBRs does not provide substantial grounds for concluding that the risks of operating the demonstration reactor and the FFTF are accepable. We have had three small LMFBRs in the United States—the Experimental Breeder Reactors I and II (EBR-I and EBR-II) and Fermi. The core of EBR-I melted down and almost exploded, due to an autocatalytic reactivity effect; Fermi suffered a meltdown of two fuel rod bundles, due to coolant flow blockage of two fuel ducts, and, further, the reactivity anomolously began to rise before the reactor was finally shut down manually eleven minutes after an alarm (see app. 2). EBR-II was to be operated as a power plant, but its emphasis was changed to that of testing fuel; it has operated rather successfully since it began in 1964, but evidently only on the average of about one half of its design power level,[101] which raises a question as to its value in providing confidence that the demonstration plant can be operated safely. Also, a random check of the EBR-II operation shows that two SCRAMs occurred in ten days, which raises further questions.[102] More importantly, the fuel and many other core design features and performance demands are very different in EBR-II.

The particular autocatalytic effect which caused the EBR-I

core meltdown accident resulted from relatively slight inward bowing of fuel rods during their heat-up, which raised the reactivity, since it tended to compact the core. The reactivity rise caused the power level to rise faster and, hence, the fuel rods to heat up faster and bow even more (see app. 2). Steps were taken in the design of the EBR-II and Fermi LMFBRs to preclude the effect by mechanically preventing significant inward fuel-rod bowing. This measure went a long way toward making these reactors stable.[103] Yet, the FFTF and LMFBR Demonstration Plant reactors are now being designed without the EBR-II and Fermi type of mechanical restraint (inward-bowing preventer), so that substantial positive bowing reactivity feedback will be allowed.[104] This positive reactivity effect is expected (predicted) to be counteracted, however, by the negative feedback due to fuel heating (Doppler effect), so that there will be no net autocatalysis, at least if the reactor is operated within prescribed ranges of temperature and coolant flow during reactor heat-up (start-up).

But it should be asked whether the thermal-mechanical bowing could possibly be underpredicted, so that an EBR-I type core meltdown event could recur. The core assembly is a complex mechanical structure, whose thermal-mechanical contortions* during heat-up are further complicated by complex temperature differences and nonuniform "swelling" of the steel ducts caused by the intense neutron irradiation. Therefore, any design prediction should be considered less than a certainty until it is borne out by actual reactor operation. The way to avoid the uncertainty would be to use the EBR-II/Fermi method of inward bowing prevention in LMFBR core designs and thus preclude the effect altogether. This method consists of placing spacer pads or buttons between the individual fuel-rod-bundle ducts at the midheight level in the core. (The core consists of many ducts packed together, except for a slight clearance between adjacent flat duct walls.) The buttons create mechanical contact beween fuel ducts (no clearance or gap) at the midheight level, so that when the ducts try to bow inward, toward the core vertical center line, they will actually push against each other, causing them all to move radially outward, effecting core expansion, and thereby reducing the reactivity, which avoids autocatalysis.

However, large LMFBR power reactors, as in the LMFBR Demonstration Plant, are expected to experience swelling of the

* Small geometric contortions of the fuel assembly (fuel-bundle ducts) in the core, barely perceptible to the eye, can cause serious reactivity increases, as in the EBR-I meltdown (CRBR PSAR: ch. 4; p. 4.2-128).

steel ducts due to neutron radiation, and this would preclude the use of the spacer buttons *in the core,* for unless there is sufficient initial clearance between the fuel ducts (and any spacers) to allow for swelling, the ducts will become stuck together with irradiation and prevent reactor refueling (removal and replacement of spent fuel bundles).[105] For this reason, the spacer buttons are placed on the ducts *above* the core section (fuel zone) and thus away from high swelling (see fig. 12). (They are still needed to keep the ducts from being loose, else the reactivity would rise dangerously when the ducts shift about. Incidentally, this raises another accident possibility, that of spacer buttons being incorrectly made so as to make the core fuel assembly loose.)

Therefore, we see that the core designs of the newer U.S. LMFBRs are a departure from the EBR-II design relative to fuel-bowing—a design which worked in preventing a recurrence of core meltdown due to autocatalytic bowing. Such departures should be examined very closely. In this regard, this author has not yet evaluated the new LMFBR design as to its adequacy, for that would be part of the needed detailed evaluation of accident prevention. The new designs may be reasonably safe relative to bowing; but the departure should at least be noted, in view of the facts that one LMFBR melted down due to the phenomenon and that the reactivity processes which the bowing triggered to produce the EBR-I meltdown accident are far from being understood (see app. 2, no. 1). It should also be noted that France's LMFBR demonstration reactor (Phénix) uses spacer buttons placed in the core, as in EBR-II and Fermi, so the fact that its operation relative to bowing is so far stable does not mean that the FFTF and the Clinch River LMFBR Demonstration reactors will be.[106]

Another question affecting *accident prevention* arises in regard to a postulated process called fuel failure propagation—that is, the failure of a fuel rod or rods, due to overheating, the effects of which cause failure of adjacent rods, and so on through the core in a cascade manner. Could such a process happen too quickly to be detected in time to shut down the reactor before a nuclear runaway occurs? Will it require whole core testing to adequately verify any predictions concerning the question? Will the LMFBR Demonstration Plant be, in effect, the experiment? These questions too should be carefully examined. For a general outline of such LMFBR safety questions, see "Safety Problems of Liquid-Metal-Cooled Fast Breeder Reactors" by C. N. Kelber et al., of Argonne National Laboratory (ANL-7657); and *The LMFBR Program Plan,* vol. 10, on safety (WASH-1110, Aug. 1968), Argonne National Laboratory.

Finally, there is the possibility that the LMFBR demonstration plant will be designed with extra safety measures with a view to "demonstrating safe operation" but that the thousand or so subsequent LMFBRs will be trimmed of these measures to save costs. The cost-benefit feasibility analysis of the AEC for the "advanced" LMFBRs given in the *Proposed Final Environmental Statement* is vague with respect to assumptions of reactor design parameters.[107]

Other Considerations. There are other LMFBR safety considerations that have not been explored in this book, such as a core meltdown without catastrophic explosion,* which should receive careful attention. And there are alternative breeder reactors to consider, the gas-cooled, fast neutron reactor, for one;[108] but since this reactor has many of the same conceivable autocatalytic reactivity effects of the LMFBR, it would have to receive the same high level of scrutiny. (The LMFBR is treated in this book because it has been assigned the priority of the AEC.) To be sure, there are disadvantages that would be incurred in addition to the advantages offered by this type of reactor; for example, the gas-cooled breeder has its own peculiar modes of core compaction.

* The core melting through the containment bottom would contaminate the ground water below.

Eleven
Remarks on Sabotage and Other Considerations

A NUMBER of important questions naturally arise in connection with nuclear reactor accident hazards. These relate to sabotage, proposals for "nuclear energy centers" (clusters of reactors), underground placement of reactors, alternate reactor design concepts, and ways to make nuclear reactors safe. The following remarks on these subjects are offered.

Sabotage. The risk of sabotage resulting in a disastrous reactor explosion is not remote. The basic information to enable a saboteur to cause a disaster is already available in the safety analysis reports and the open technical literature, and of course, a reactor plant employee could trigger an explosion much more easily. Therefore, the risk of sabotage must be assessed; but this cannot be done with vague, general references to the problem and official assurances that security measures are strong and that sabotage would be very difficult.[1] Rather, one needs to consider several *specific* possibilities for sabotage in order to assess its potential ease of accomplishment and, if nuclear energy is to be developed further, the extent of the safeguards that would be necessary. For example, a *multiple control rod ejection* accident (an extremely severe power excursion) could conceivably be triggered by a timed dynamite bomb placed among the control rod drive mechanisms on the reactor vessel in either a PWR or a BWR. The dynamite explosion could sever several mechanism housings, which would cause the ejection of the affected control rods by the high-pressure reactor coolant and, in the case of a BWR, could also destroy simultaneously, and thereby negate, the safety protection against control rod ejections, namely, the control rod movement blocking structure (see figs. 1 and 2). Other possibilities for sabotage suggest themselves from the discussion in the previous chapters and from available safety analysis reports, including the Rasmussen Report. Indeed, sabotage was specifically included in the "Enumeration and Evaluation of Possible Hazards" in the *Hazards Evaluation Report* for the experimental boiling water reactor (EBWR). After discussing several sabotage possibilities, the report concluded: "It is not possible to incorporate safety features consistent with the aims of a dedicated saboteur." [2] In regard to present nuclear power plants, it should be noted that Professor Rasmussen has publicly revealed that the Sandia Laboratory has prepared a "classified study" of "exactly how" a saboteur could bring about a disastrous "accident," including the maximum accident.[3] To be sure, plant security will be strong, but whether this will be adequate is another of the subjective judgments that will have to be made.

Nuclear Energy Centers. Consideration is being given by the U.S. Nuclear Regulatory Commission, as required by the Energy Reorganization Act of 1974,[4] to the building of ten to forty large nuclear power reactors at a single site.[5] About ten reactors would be located within a mile of each other. One main advantage of such a concept is that the spent fuel rod bundles can be reprocessed on the site to recover the plutonium fuel for recycling in the reactors, thereby avoiding the transportation of the plutonium off-site, where the chances may be greater for theft and subsequent use of the stolen plutonium in making atomic bombs. However, such clustering of reactors raises conceivable possibilities for colossal multiple-reactor accidents. For example, an earthquake that might cause an accident in one reactor would presumably cause the same accident in adjacent reactors, due to a common design. Or, a disastrous reactor accident in which one half of the core and its fission products are ejected out of the reactor containment—as the Rasmussen Report admits is possible even for the water-cooled reactor—would create such heavy radiation in the region of the site that the operators of the remaining reactors might flee, justifiably, leaving their reactors unattended. (The direct rays of radiation from the fallen radioactive debris would penetrate over a mile in all directions.[6])

The possibility for common-mode or chain-reaction accidents presently exists in such twin-unit reactor sites as the Browns Ferry complex, which has already suffered a serious accident (fire) that caused the shutdown of both reactors simultaneously (see appendix), or the three-reactor Indian Point complex in New York, near New York City.

Underground Placement. It is often suggested that reactors be built underground, as a way to avoid a heavy release of radioactivity into the atmosphere upon an accident. While this is easy to suggest, it is evidently not practical, economically or technically. Problems of flooding or collapse of the reactor plant cavities, added costs, and adequate sealing arise. At best, it would take about ten years to design and build a prototype underground containment system to gain the necessary practical experience, assuming engineering studies would show that underground placement is feasible. However, according to the AEC, no definitive engineering studies even exist.[7]

Alternate Reactors. Of special interest is the CANDU reactor, which is widely in use in Canada.[8] This reactor has several safety advantages; mainly, it does not require the successful operation of its emergency core cooling system to avoid a containment rupture in the event of a loss-of-coolant accident,[9] because the reactor design is such that no fuel melting would occur. However, an autocatalytic power excursion accident would definitely occur during a loss-of-coolant accident, if a reactor SCRAM fails to occur, which would cause a core meltdown. (The CANDU has a backup SCRAM system.) It cannot be demonstrated theoretically that the containment would not be breached in this event.[10] Since the fission product radioactivity content is only slightly less than the U.S. reactor designs, the consequences of such an accident in CANDU could be disastrous, and therefore, the CANDU must be more closely examined.

There are other reactor concepts which should be mentioned, namely, the high-temperature, gas-cooled (HTGR) and the light water breeder reactor (LWBR).[11] A 330-Mw HTGR is being operated in Colorado and construction permits for four larger HTGRs (about 1000 Mw) are being sought.[12] The HTGR uses uranium fuel embedded in graphite blocks pierced with holes for cooling by helium gas. The entire reactor system is contained in a reinforced concrete pressure vessel, which in turn is contained in a containment building. Thus, the HTGR is completely different than water-cooled reactors and has different accident modes. There are some advantages relative to safety, such as no autocatalytic reactivity effects due to coolant changes. (The HTGR will ultimately depend on a plutonium-fueled, fast neutron breeder for fuel, namely plutonium.) However, as in a water-cooled reactor, the core, to avoid overpressure, needs to be kept from overheating by continuous cooling. Also, a *Nuclear Safety* article on the "Safety-Design Basis of the HTGR" does not analyze the possible accidents that would be more severe than the design basis accidents. Of concern are the possibility of a control rod ejection accident and, in the event of loss of cooling and rupture of the containment, the possibility of igniting the graphite (carbon), as in the Windscale reactor accident in England (see app. 2, no. 14), thereby causing a near full release of strontium and cerium fission products. Clearly, a full hazards analysis must be developed for the HTGR, especially since a large plant is in operation.

The LWBR is of no serious interest at the moment, except for considering it as a possible alternative nuclear power reactor. A demonstration reactor for the LWBR is being built into the Shippingport facility near Pittsburgh, Pennsylvania, but it will be operated at only 55 Mw, or 5% of the power rating of large nuclear plants. At worst, the demonstration reactor could have accident consequences approaching the WASH-740 estimates, but since these are serious enough, it too needs to be examined.

Can anything be done to make reactors safe? As the preceding chapters should show, this question does not admit an answer, since safety is subjective. Certainly, we would have to resolve much of the scientific uncertainty before we could judge what types of added safety features would provide the degree of extra precaution society may desire. However we may already be at the limit of complexity in providing safety equipment; any more features may detract from the present level of safety, besides driving up the costs.

A final note: the accident hazards of naval reactor plants should be thoroughly examined as well, since the navy's plans for "long-life" reactor cores will mean that each naval core (for the *Nimitz*-type aircraft carrier) will build up a great amount of strontium 90 radioactivity—at least 25% of the strontium 90 content in a large, 1,000-Mw, civilian power plant,[13] whose potential accident consequences were estimated in chapter 1.

Twelve

Conclusions

THE ACCIDENT risk of present-day nuclear power plants, namely, the water-cooled reactors, has not been scientifically established. The reactor plants are designed to control only a limited set of accidents, *design basis accidents* (DBAs); but the calculations behind the design predictions are based on experimentally unverified theory and arbitrary assumptions. To remove these shortcomings would require a sizable experimental program involving large or full-scale reactors, as present facilities are not adequate. Moreover, there are many possible accidents, worse than the DBAs, that appear to have a potential for severe reactor explosion and consequent radioactivity release. These have not been adequately investigated by the nuclear community. Furthermore, to experimentally investigate the explosion and radioactivity release potential of these *worse possible accidents* (WPAs) appears to be impractical, other than to perform several bounding, worst-conceivable-nuclear-runaway and "'core meltdown" experiments. Without such bounding-type destructive tests, the hazards of the WPAs will never be known, but we may assume a near full release of fission products, which could have intolerable consequences, such as the contamination of agriculture over many large states. Finally, a judgment that the probability of accidents (DBAs or WPAs) is negligible or acceptably low is extremely subjective; therefore, any numerical accident probability estimates are unreliable and no valid basis for dismissing the WPAs. In order to make a rational judgment, a full analysis must be performed of both the DBA and WPA hazards and associated uncertainties and of the chain of failures needed to cause each accident, along with the related history of actual reactor malfunction and mishaps.

The hazards and uncertainties of the LMFBR are even more serious due to a *nuclear explosion* potential. The maximum LMFBR explosion potential has not been scientifically established to be safely containable for the DBAs, nor is the containment designed to withstand real accident possibilities more severe than the DBAs. Disastrous nuclear explosions are theoretically conceivable even for the DBAs (possibly 150,000 square miles of land requiring abandonment due to plutonium and fission product contamination); and there presently is no theoretical bound on the explosion potential. A great deal of additional theoretical analysis is required. Furthermore, it is very doubtful that experiments for verifying the theoretical accident calculations are practical, again, even for the DBAs, because of the extreme complexity of the physics of LMFBR fuel-meltdown accidents.

These findings call for a full investigation and exposition of the full accident potential of reactors and associated uncertainties, including all possible and conceivable accidents and related reactor experience, and of the need for experimental verification of theory, with particular attention to ensuring objectivity, so that society can make a sound judgment as to the course the nation should follow.

In the meantime, in view of the enormous potential for disaster, it is difficult to escape a conclusion that civilian nuclear reactors ought to be shut down* and a moratorium placed on nuclear plant construction, while society resolves these issues and determines the safe course of action. Furthermore, the large-scale LMFBR demonstration reactor project presently underway in Tennessee (the Clinch River Breeder Reactor) as well as the plutonium-fueled, LMFBR-like Fast Flux Test Facility reactor (FFTF) about to be operated in Washington should be postponed indefinitely, so that the ill-defined risks of geographically widespread plutonium and fission product radioactivity contamination may be avoided while society determines the proper course of action.

Unfortunately, shutting down the nation's nuclear plants is fraught with difficulty, for about 9% of the nation's electrical energy is supplied by nuclear power plants.[1] Indeed, about 30% of Chicago's Commonwealth Edison's electrical capacity is nuclear.[2]

* Merely reducing the operating power level of the reactors will not substantially reduce the accident risk, since many worst conceivable power excursion accidents would occur when the reactor is at zero power, due to the nature of reactivity accidents.

Obviously, shutting down the reactors would cause a serious hardship, but of course, the effects of a reactor accident could be much worse. In view of the hazards and uncertainty, it would seem prudent to shut down the plants, rather than risk the health and safety of not only those people served by nearby nuclear plants but those in surrounding areas (involving several states) as well. True, Chicago, for example, would especially suffer from the lack of electrical energy, but must the people downwind accept the risk of accidents so that the Chicago area can operate its nuclear plants? Presumably, the Chicago area could receive electricity from outside nonnuclear plants through the electrical power grids, until the necessary adjustments could be made. On the other hand, it may be totally impractical, from the standpoint of local economies, to shut down the reactors when such possible consequences as increased unemployment, crime, and other health or safety hazards are considered. A difficult decision is ahead, indeed.

There is one final consideration. Inasmuch as the water-cooled reactors will depend on the LMFBR for fuel (by the breeding process) after twenty-five years or so, according to present plans, the LMFBR explosion hazard issue must be resolved concurrently with the water-reactor safety issue. Otherwise, society would have to assume that the LMFBRs will be judged unsafe and consider the water-cooled reactors as only a short-term energy option. A determination that the LMFBR is unsafe would affect a judgment on the safety of water-cooled reactors, since it would have to be decided whether the short-range benefits of operating the reactors outweigh the accident and other risks, including the long-term radioactive waste disposal problem that would also be left behind.

Thirteen

Who Should Decide?

AFTER a thorough review of the safety of civilian nuclear power reactors is conducted, a fundamental decision will have to be made: whether to (a) proceed with the nuclear energy program and accept the risks; (b) postpone the development and use of nuclear reactors and perform further hazards research, deferring a final decision until the results are evaluated; or (c) reject the use of nuclear energy. This chapter poses the question, Who should decide this issue? That is, who should decide whether nuclear reactors are safe?

The federal government, namely, Congress, has assumed the authority to promote nuclear energy by the passage of the Atomic Energy Act of 1954 and its subsequent amendments, including the Energy Reorganization Act of 1974, thereby making the basic judgment that nuclear reactors are safe, or can be made safe in a practical way.[1] By and under these acts, an elaborate administrative structure has been established for assessing the hazards of reactors and for deciding the safety issues and research needs. This structure includes the Nuclear Regulatory Commission, the Advisory Committee on Reactor Safeguards, the Atomic Safety and Licensing Board, and the Joint Committee on Atomic Energy of the Congress. But will this system of decision-making best ensure our safety and well-being? Can we trust that this system will lead to a sound decision? Or should the basic decision of safety not be left to the experts of the nuclear agencies and laboratories, or to the Joint Committee, or even to Congress? Should the basic decisions instead be referred to the States, which are closer to the people, or to the people themselves? This author contends that it is essential that these questions be taken up and, further, that an appeal be made to the fundamental constitutional principles of the land, for reasons which will be given next.

As one can appreciate from the analysis of reactor accident hazards given in the preceding chapters, a determination of the safety of nuclear reactors will depend primarily on the *subjective* judgments and personal philosophies of the particular set of individuals who will hold the power to decide the reactor safety issue. Briefly, the subjective elements include: the acceptability of unverified theory for predicting the course of postulated accidents; the rejection of the experimental philosophy; the subjective judgments factored into the theories of reactor explosion accidents; the assumed levels of acceptable radiation exposure to the population following an accident, which set the contamination limits for estimating possible accident consequences; the likelihood of accidents; the need to calculate the force and potential consequences of the worst possible accidents; and the practicality, safety, and value of integral reactor destructive experiments. To show that a theory for predicting the course of possible accidents has not been verified experimentally, and that disastrous accidents are conceivable, would not necessarily persuade the decision makers to postpone the development of nuclear energy, or to abandon it, if these persons hold the view, as does the U.S. Nuclear Regulatory Commission, for example, that the theory is basically adequate and that the likelihood of serious accident is small enough so that the benefits of nuclear power warrant the risks.[2] Presently, these reactor safety judgments are left largely to the experts in the federal government, who are necessarily nuclear engineers and scientists and who thus have a natural vested interest in sustaining nuclear power, though they try, of course, to develop technological solutions to safety problems, mainly accident preventive measures. This vested interest will naturally influence the judgments of the federal regulatory agencies. Indeed, the Atomic Energy Act mandates to these agencies that nuclear energy be developed "so as to make the maximum contribution to the general welfare." Clearly, therefore, a judgment on the safety of the people relative to nuclear energy will depend on *who decides.*

Intertwined with this question of *safety* is the question of *necessity:* does our society need nuclear energy? Here again, human value judgments will surely control the answer. Though not the subject of analysis in this book, the economic and security considerations that will factor in a judgment on the necessity of nuclear energy will obviously depend on subjective assessments of the economic risks of nuclear energy, on the way of life society may desire or find necessary for ecological and economic reasons, and on the desired level of national security, which may or may not require civilian nuclear energy, depending on personal judgment. There are also the health hazards and environmental impact of using fossil fuels to consider; nuclear energy could avoid these risks, at least for the power-plant phase of energy production. It is not the purpose of this present analysis to delve into these "benefit" considerations, except to say that surely an appraisal of the benefits or necessity of nuclear energy will depend heavily on how the risks of reactor accidents are judged. (There are other hazards of nuclear energy, which have not been examined in the present analysis, such as radioactive waste seepage, and theft of nuclear fuel by terrorists for use in fashioning crude atomic bombs. Of course, such considerations will also figure in the risk-benefit judgment.)

It is rather obvious in light of the value of nuclear power as an energy resource and of the hazards and countering safety measures, which are difficult to assess, that the nuclear reactor safety issue is one of the most profound and difficult decisions on human values facing society. It would be both wise and appropriate at this critical juncture, therefore, to reflect on the fundamental, *constitutional* principles of the land that bear on this issue, to determine where the People, by their constitutions, have vested the authority for resolving it. For in a way of thinking, our state and federal constitutions were established to ensure that society make sound human value judgments by carefully prescribing the powers of government in regard to *who* shall make *what* kinds of value judgments. This follows from the Declaration of Independence, which sets forth the fundamental axiom on which the constitutions were founded, to wit:

> Governments are instituted among Men, deriving their just powers from the consent of the governed, —That . . . it is the Right of the People . . . to institute new Government, laying its foundation on such principles and organizing its powers in such form, as to them shall seem most likely to effect their Safety and Happiness.

This political axiom finds expression in the Constitution of the United States—the supreme Law of the Land—in the form of the Tenth Amendment:

> The powers not delegated to the United States by the

Constitution, nor prohibited by it to the States, are reserved to the States respectively, or to the people.

From the above principles, we can and must assume that the method implied by the present Constitution for resolving the nuclear safety issue is the wisest and only proper method, until and if the Constitution is changed.

This author has pursued this constitutional question in a rigorous study of constitutional law and history and would like to offer his interpretation of the Constitution in this regard. It is his opinion that the nuclear program is unconstitutional. Specifically, it is contended that the United States Constitution does *not* delegate to the federal government (Congress) powers which can be reasonably construed as authority to promote and regulate civilian nuclear power plants—namely, powers to spend money from the treasury to develop and subsidize nuclear reactors; to regulate their operation and location, and to preempt any laws of the states that would require more stringent regulations or prohibition; and to grant the nuclear industry immunity from liability in the event of a severe reactor accident, as has been done by the Price-Anderson amendment to the Atomic Energy Act.[3]

The foregoing interpretation of the Constitution, if correct, would answer immediately the quesion of who should decide the nuclear safety issue, and by what process. For the answer would follow from Article V of the Constitution, which defines the process for amending the Constitution. Applying Article V, Congress would have to review the safety of nuclear reactors; and if Congress should conclude that nuclear energy is safe and that the federal nuclear program should continue, then it would have to ask the People for the authority to promote and regulate nuclear energy, or some broader authority, by proposing a constitutional amendment to the states. The states by considering an amendment proposition would then make the basic human value judgment as to the safety and environmental risks of nuclear reactors—unless Congress should repeal the nuclear program following its review.

Amendments may be ratified either by the legislatures of the states or by state conventions.[4] The latter mode may be preferable in that the people would elect delegates specifically selected for the purpose of deciding the profound nuclear safety issue. In either case, ratification by three fourths of the states would be necessary for passage of an amendment, before the federal government could continue with its civilian nuclear energy program.

We can see that if the nuclear program were judged unconstitutional the question of whether the present administrative decision-making structure within the federal government is optimum or adequate to judge the safety issue would become moot, as the basic decision on nuclear safety would have to be made by a wholly different process, outside the administrative structure, designed to effect a closer consultation with and a thorough review by the People. Included among the People, incidentally, is the scientific community at large, whose attention would be secured by this decision process and who would surely contribute to the making of a sound public decision, not only by reviewing analyses of reactor safety but by taking up the fundamental issue of the experimental philosophy—the need for full-scale reactor testing—and giving it a thorough consideration. These are the reasons that the constitutional inquiry is fundamentally important and must be taken up.

A brief legal argument in support of the foregoing opinion on the Constitution is offered below, to show that there is a serious question of the constitutionality of the federal civilian nuclear program. It is included in this treatise on reactor accident hazards, rather than in a separate work, because, as we shall see, the specific legal assumptions being challenged are so commonly accepted, and have such far-reaching implications, that to merely question them without supporting argument might be met with indifference. But unless the constitutional question is taken up, the nation stands the risk of not making a sound decision on nuclear safety, because the field of inquiry into *who should decide* will have been limited.

In addition, since the safety of nuclear reactors is as much a question of *who decides* as it is a question of the scientific uncertainties and accident possibilities and probablities, it is important for those who take the time to learn and grasp the subjective aspects of assessing the accident hazards to consider the importance of a constitutional inquiry as well and, by examining the legal argument below, to appraise the validity of the constitutional question being raised. Those who comprehend the full, subjective character of nuclear safety and who at the same time comprehend the constitutional uncertainty may better appreciate the need to take a broad view of the question of who should decide, instead of restricting it to the federal administrative structure.

And since time is of the essence—because of the risks of accidents that already exist and because the construction of new reactors has been accelerating—it is felt that both the *hazards* and the *decision process* should be considered urgently, if we are to ensure a sound public decision, which is, after all, what we all are after.

For the foregoing reasons, the following brief legal argument is offered.

Constitutional Analysis and Argument

We begin by reviewing the powers of Congress. These are basically specified in Article I, section 8, of the U.S. Constitution:

> The Congress shall have Power To lay and collect Taxes, Duties, Imposts and Excises, to pay the Debts and provide for the common Defence and general Welfare of the United States; but all Duties, Imposts and Excises shall be uniform throughout the United States;
>
> To borrow Money on the credit of the United States;
>
> To regulate Commerce with foreign Nations, and among the several States, and with the Indian Tribes; . . .
>
> To coin Money, regulate the Value thereof, and of foreign Coin, and fix the Standard of Weights and Measures; . . .
>
> To establish Post Offices and post Roads;
>
> To promote the Progress of Science and useful Arts, by securing for limited Times to Authors and Inventors the exclusive Right to their respective Writings and Discoveries; . . .
>
> To declare War, grant Letters of Marque and Reprisal, and make Rules concerning Captures on Land and Water;
>
> To raise and support Armies, but no Appropriation of Money to that Use shall be for a longer Term than two Years;
>
> To provide and maintain a Navy; . . .
>
> To provide for calling forth the Militia to execute the Laws of the Union, suppress Insurrections and repel Invasions; . . .
>
> To exercise exclusive Legislation in all Cases whatsoever, over . . . the Seat of the Government of the United States, and to exercise like Authority over . . . Arsenals, dock-Yards, and other needful Buildings;—And
>
> To make all Laws which shall be necessary and proper for carrying into Execution the foregoing Powers, and all other Powers vested by this Constitution in the Government of the United States, or in any Department or Officer thereof.

A review of the Atomic Energy Act and its legislative history [5] reveals that the civilian nuclear power program is based on the express assumption that Congress has the indefinite power *to provide for the general welfare* and the implicit assumption that

Congress has a substantive power *to promote and regulate manufactures,* that is, the manufacturing of electricity by nuclear means. These assumptions are allegedly grounded on the *taxation* clause, particularly its common-defense and general-welfare subclause (hereafter referred to as the *welfare clause*), and on the *commerce clause,* which are the first and third clauses listed above. To show that this is the case, we need only refer to chapter 1 of the Atomic Energy Act, which asserts the "Findings" of Congress:

> . . . the processing and utilization [power plants] of . . . nuclear [fuel] material affect interstate and foreign commerce and must be regulated in the national interest . . . and in order to provide for the common defense and security and to protect the health and safety of the public. . . .
>
> Funds of the United States may be provided for the development and use of atomic energy under conditions which will provide for the common defense and security and promote the general welfare.
>
> In order to protect the public and to encourage the development of the atomic energy industry, in the interest of the general welfare and of the common defense and security, the United States may make funds available for a portion of the damages suffered by the public from nuclear incidents, and may limit the liability of those persons liable for such losses.

However, the Congress simply claimed such powers without offering any historical or legal analysis. It is contended that the *welfare clause* and *commerce clause* confer no such powers. We shall examine first the welfare clause.

Clearly, a power to provide for the general welfare is *indefinite* and therefore is virtually an unlimited power, depending on the opinion of Congress as to what is needed for the general welfare; whereas the powers enumerated in Article I, Section 8, are *specific.* However, since the alleged power to provide for the common defense and general welfare would include the specific powers by implication, the latter would be superfluous. It seems contradictory that the Constitution should go to great lengths to specify particular powers, only to make them superfluous by a sweeping power to provide for the general welfare. And why should such a paramount, sweeping power be defined in a manner that appears, grammatically, to be only a qualification of the power to tax, as indicated by the system of punctuation and capitalization, which relegates the welfare clause to second status at best. It would have to be admitted that there is a possibility that the welfare clause is only a qualification of the power *to tax,* not a grant of power in itself, such as an indefinite power to spend money for the "general welfare." Also, nuclear power plants could be considered an activity of *manufacture,* not of *commerce;* so that there would be no power stemming from the commerce clause to regulate nuclear plants, much less to promote their development.

Since the Constitution is subject to interpretation, it is necessary to settle on the proper method of interpreting written laws—the Constitution being the fundamental law. Said Blackstone, whose 1765 *Commentaries on the Laws of England* was a fundamental reference on the rudiments of law for our Founding Fathers,[6] "The fairest and most rational method to interpret the will of the legislator is by exploring his intentions at the time when the law was made, by *signs* the most natural and probable." [7] The signs include the *context* of the words in the law, and the *reason* and *spirit* of the law. It is necessary, therefore, that we explore the recorded intentions of the makers of the Constitution. These will be found mainly in the records of the Federal Constitutional Convention[8] and the States' conventions that ratified the Constitution;[9] and in *The Federalist,* which was a collection of eighty-five essays written by three prominent advocates of the Constitution (Madison, Hamilton, and Jay) during the ratification debates to explain and justify the Constitution. Some highlights of these records are presented below, along with a discussion of relevant Supreme Court opinion on the subject questions.

The records of the making of the Constitution show that the *taxation* clause is no grant of power to provide for the general welfare, nor is it a power to *spend* money from the federal treasury for the broad and undefined objects of the general welfare. It is merely a power to *raise* money—a power that the Continental Congress lacked under the previous Articles of Confederation, which was one of the deficiencies that led to the convening of the Federal Constitutional Convention.[10] The spending power (namely, the money appropriations power) was assumed to be plainly implied in, and confined to, the *enumerated* powers—for example, to pay for armies, a navy, and post offices, and to pay the expenses incidental to the work of laying and collecting taxes and regulating commerce—as spending money is obviously a *necessary and proper* means to execute the specific powers.

The *welfare* clause was affixed to the *taxation* power only to define and limit the *purposes* for which taxes could be laid and collected.[11] Originally, in the Constitutional Convention, the clause "To lay and collect Taxes, Duties, Imposts and Excises" stood alone, but the framers wanted the People to understand that the tax revenues would be used to repay the debts incurred during the Revolutionary War—hence, the clause "to pay the Debts" was affixed. To leave it at that would have meant that taxes could be laid and collected only to pay debts, and so the clause "and the necessary expenses of the United States" was added. But even this was thought not to be sufficiently broad, since "Duties" on imports, for example, might be necessary, not to raise money to pay expenses of government, but to regulate commerce, as in the cases of protective and retaliatory tariffs. Also, care had to be taken to ensure against the exercise of the taxation power by Congress to the detriment of a political minority.

Hence, the framers qualified the taxing power so that it could be exercised only for the common defense and general welfare (and for paying the debts). For the welfare-clause phraseology, they simply fell back on the eighth of the Articles of Confederation, to wit:

> All charges of war, and all other expenses that shall be incurred for the common defence or general welfare, and allowed by the united states in congress assembled, shall be defrayed out of a common treasury which shall be supplied by the several states.[12]

This article contained no grant of power to *spend* money; rather, it merely established a treasury and explained how it was to be plenished. The ninth article of the Confederation contained the enumeration of the specific powers of the Continental Congress. Thus, just as the *welfare* clause in the Articles of Confederation was well understood by the framers of the Constitution to be no grant of power to spend money for the indefinite general welfare, so, too, the *welfare* clause in the Constitution was to be no such grant of power.

The Federalist (McLean ed., 1788, essay no. 41) specifically addressed the question of whether the welfare clause is a grant of power and explained at length that it definitely is not:

> Some who have not denied the necessity of the power of taxation have grounded a very fierce attack against the Constitution, on the language in which it [the power of taxation] is defined. It has been urged and echoed, that the power "to lay and collect taxes, duties, imposts and excises, to pay the debts, and provide for the common defense and general welfare of the United States," amounts to an unlimited commission to exercise every power which may be alleged to be necessary for the common defense or general welfare. No

stronger proof could be given of the distress under which these writers labor for objections, than their stooping to such a misconstruction.

Had no other enumeration or definition of the powers of the Congress been found in the Constitution than the general expression just cited, the authors of the objection might have had some color for it; though it would have been difficult to find a reason for so awkward a form of describing an authority to legislate in all possible cases . . . by the terms "to raise money for the general welfare."

But what color can the objection have, when a specification of the objects alluded to by these general terms immediately follows and is not even separated by a longer pause than a semicolon? . . . For what purpose could the enumeration of particular powers be inserted, if these and all others were meant to be included in the preceding general power?

In the Virginia ratification convention the opponents to the Constitution feared that the welfare clause would eventually be construed by Congress as a grant of power. The Virginia governor, John Randolph, who was also a delegate to the Federal Constitutional Convention and who advocated the Constitution in the Virginia convention, assured the delegates that the welfare clause was no grant of power:

> But the rhetoric of the gentleman has highly colored the dangers of giving the general government an indefinite power of providing for the general welfare. I contend that no such power is given. They [Congress] have power "to lay and collect taxes, duties, imposts, and excises, to pay the debts and provide for the common defence and general welfare of the United States." Is this an independent, separate, substantive power to provide for the general welfare of the United States? No, sir. They can lay and collect taxes, &c. For what? To pay the debts and provide for the general welfare. Were not this the case, the following part of the clause would be absurd [i.e., "but all duties, imposts, and excises shall be uniform throughout the United States"]. It would have been treason against common language. Take it altogether, and let me ask if the plain interpretation be not this—a power to lay and collect taxes, &c., in order to provide for the general welfare and pay debts.[13]

Near the end of the Virginia convention when the opponents of the Constitution sensed defeat, they reverted to worrying about the welfare clause. Randolph laid the issue to rest when he said that they were

> back to the clause giving that dreadful power, for the general welfare. Pardon me, if I remind you of the true state of that business. I appeal to the candor of the honorable gentleman, and if he thinks it an improper appeal, I ask the gentleman here, whether there be a general, indefinite power of providing for the general welfare? The power is, "to lay and collect taxes, duties, imposts, and excises, to pay the debts, and provide for the common defence and general welfare"; so that they [Congress] can only raise money by these means, in order to provide for the general welfare. No man who reads it can say it is general, as the honorable gentleman represents it. You must violate every rule of construction and common sense, if you sever it from the power of *raising* money, and annex it to any thing else, in order to make it that formidable power which it is represented to be.[14]

Most of the state conventions attached numerous reservations to their ratification resolutions, which called for further restrictions on the power of Congress, not enlargement. These resolutions led to the Bill of Rights, *and the Tenth Amendment,* quoted earlier,[15] which emphasizes that Congress has no powers except those delegated in the Constitution.

Furthermore, in the Federal Constitutional Convention it was proposed to add to the powers of Congress, the powers "To establish public institutions, rewards and immunities for the promotion of agriculture, commerce, trades, and manufactures"; and "To encourage, by premiums and provisions, the advancement of useful knowledge and discoveries."[16] But these powers, which would have allowed the civilian nuclear power program, were rejected.

Note that Article I, section 8, of the Constitution includes the power "To promote the Progress of Science and useful Arts, by securing for limited Times to Authors and Inventors the exclusive Right to their respective Writings and Discoveries." Now, if the Founding Fathers wanted Congress to have a power to promote science and useful arts (technology) by providing money for research and development, then surely they would have left off the qualifier, "by securing for limited Times to Authors and Inventors the exclusive Right to their respective Writings and Discoveries," which limits the granted power *to promote science and useful arts* to providing for patents and copyrights, and that only.

The Federalist explained the limited powers of Congress in relation to the States:

> In the first place it is to be remembered that the general [federal] government is not to be charged with the whole power of making and administering laws. Its jurisdiction is limited to certain enumerated objects. . . .
>
> The powers delegated by the proposed Constitution to the federal government are few and defined. Those which are to remain in the State governments are numerous and indefinite. The former will be exercised principally on external objects, as war, peace, negotiation, and foreign commerce; with which last the power of taxation will, for the most part, be connected. The powers reserved to the several States will extend to all the objects which, in the ordinary course of affairs, concern the lives, liberties, and properties of the people, and the internal order, improvement, and prosperity of the States.[17]

This principle of defined, limited, enumerated powers would be nonexistent if the People had granted Congress the *indefinite* power of providing for the general welfare.

In 1906 the U.S. Supreme Court had its first occasion to address the welfare clause. In *Kansas* v. *Colorado* it ruled that Congress has no indefinite power to undertake projects of internal improvements, such as "reclamation of arid lands," on the claim of "a supposed general welfare." The court held that the Tenth Amendment precluded such power.[18] Indeed, as late as 1924 the U.S. Senate issued its official analysis of the Constitution, which stated:

> The general welfare clause contains no provision of power, of itself, to enact any legislation, but on the contrary, the words "and provide for the common defense and general welfare" is a limitation of the taxing power of the United States, and that only.[19]

In 1935 the U.S. Supreme Court in *U.S.* v. *Butler* held that the welfare clause conferred no power to the federal government to regulate agriculture and ruled unconstitutional the Agricultural Adjustment Act, which imposed certain taxes on mills and subsidized selected farming under contracts with individual farmers to limit production and accept other regulations of agriculture.[20] Fairly interpreted, this *Butler* opinion implies that the civilian nuclear program cannot be founded on the *welfare clause*.

However, the court in its *Butler* opinion attempted to settle a separate question of the *welfare clause* that was not before the court as an issue for adjudication.[21] In an *obiter dictum* the court asserted that Congress can spend money for the indefinite objects of the general welfare, provided no "contractual obligations" are involved.[22] This dictum seems to have been a signal that the court was willing to accept as constitutional the new social security program, but because of the qualifier, it would still not allow the federal civilian nuclear program. In subsequent cases the Supreme Court held that the federal spending programs of social security and aid-to-housing are constitutional.[23] In these cases, however, the court would not consider the welfare-clause issue, for it was considered settled by the Butler dictum.[24] Then, in 1950, the Supreme Court in *U.S.* v. *Gerlach Livestock Co.* dropped the *Butler* no-contracts qualifier and opined that the welfare clause delegates to Congress "a substantive power to tax and appropriate [money] for the general welfare," including a power "to promote the general welfare through large-scale projects for reclamation, irrigation, or other internal improvement," [25] which conflicts with the court ruling in *Kansas* v. *Colorado.*

As in the Butler case, the *Gerlach* opinion on the welfare clause was *obiter dictum,* since it was an attempt to settle a question not presented to the court. The dictum is of questionable weight, since the court was not presented with arguments on the question by the litigants. If the quesion were submited to the court, the litigants, who would then have had a personal stake in the decision, would have taken care to research and argue the question thoroughly to illuminate the subject for the court. Moreover, the power of the judiciary extends only to *cases*, that is, controversies, submitted to the court and not to questions outside of the court.[26]

More importantly, the court in *Gerlach* offered no grounds for its opinion on the welfare clause, other than to cite the previous *Butler* dictum, while neglecting its qualifier. Thus, the power of Congress to promote civilian nuclear energy rests on the *Butler* dictum, through *Gerlach,* even though the no-contracts qualifier would seem to disallow the nuclear program. It remains, therefore, to examine the basis given by the court for its *Butler* dictum.

In support of its dictum on the welfare clause, the court in *Butler* cited only the *Commentaries on the Constitution* by Joseph Story, who expounded the view that the court adopted. Story's basic references were President James Monroe's 1822 "Message on Internal Improvements," [27] which asserted that Congress had some very limited form of spending power under the welfare clause, and Alexander Hamilton's 1791 commentary on the welfare clause. However, Monroe's message was misrepresented by

Story, since Story neglected the fact that President Monroe specifically held in his message that Congress does *not* have the power to make internal improvements (such as roads and canals) based on the welfare and commerce clauses or on any other clauses of the Constitution. (Monroe vetoed the Cumberland Road bill for that reason.) [28] Said Monroe: "I think that I am authorized to conclude that the right to make internal improvements has not been granted by the power 'to pay the debts and provide for the common defence and general welfare.' " * [29] As for his view that the welfare clause contains some limited form of spending power, such as for disaster relief, Monroe admitted that "in the more early stage of our government" he believed otherwise,[30] namely, that the welfare clause did not have even a limited spending power implied in it, which is a better indication of the intentions of the makers of the Constitution than his slightly relaxed view adopted years later. Even so, his later view would not allow the civilian nuclear program, as that definitely would come under the heading of *internal improvements* legislation, which Monroe held to be unconstitutional.

The Hamilton comment cited by Story is as follows:

> It is, therefore, of necessity, left to the discretion of the National Legislature to pronounce upon the objects which concern the general welfare, and for which, under that description, an appropriation of money is requisite and proper. And there seems to be no room for a doubt, that whatever concerns the general interests of learning, of agriculture, of manufactures, and of commerce, are within the sphere of the national councils, as far as regards an application of money.[31]

Hamilton asserted this power of appropriating money for the general welfare in his December 1791 "Report on Manufactures" as secretary of the treasury. He offered no supporting historical information for his claim. Moreover, *The Federalist,* which he coauthored, writing fifty-one of the eighty-five essays, asserts the opposite (see no. 41 by Madison). In the fifty-one essays written by Hamilton, nowhere did he make the claim later asserted in his "Report on Manufactures," nor did his essays even mention the *welfare clause.* Nor did he assert such a claim when he addressed the New York convention which ratified the U.S. Constitution.[32] If he had, the convention would have rejected the Constitution,

* Monroe denied, too, that any other clauses confer such power: "Having now examined all the powers of Congress under which the right to adopt and execute a system of internal improvement is claimed and the reasons in support of it in each instance, I think that it may fairly be concluded that such a right has not been granted."

since the States were keenly intent on limiting the powers of Congress and were repeatedly assured that the Constitution does precisely that; that is, the welfare clause is no grant of power.[33]

Indeed, very early in the Federal Constitutional Convention, Hamilton proposed a plan of Government which would have given Congress a single, unlimited power "to pass all laws whatsoever . . . ," as he suggested the virtual dissolution of the states, replacing them with one general government for the country.[34] Specifically, Article VII of his plan stated: "The Legislature of the United States shall have power to pass all laws which they shall judge necessary to the common defence and general welfare of the Union." [35] However, his plan received no support in the convention. As Hamilton himself said, "He was aware that it went beyond the ideas of most members." [36] It thus appears that Hamilton as secretary of the treasury attempted, after the Constitution was adopted by the People and after the federal government began operation, to effectuate broad, unlimited power for the federal government, for which he could not gain approval when it was established.

Finally, some have argued that the *welfare clause,* on the face of it, clearly confers a power to provide for the general welfare. If we consider the rules of English grammar known by the framers of the Constitution, however, it can be argued that the manner in which commas, semicolons, and the conjunctive word *and* are employed throughout Section 8 of Article I, plus the selected capitalization of the word *To,* have so weakened the connection of the welfare clause to the governing verb—"Congress shall have Power To"—that the clause was probably intended to be no more than a purpose of the power to tax.[37]

The preceding should be sufficient to raise the question of whether the civilian nuclear energy program can be grounded on the welfare clause.

As for the *commerce clause*—namely, the power "To regulate Commerce" among the states—the makers of the Constitution intended that the word "Commerce" be distinguished from "Manufactures," as each were to denote separate fields of human industry, as evidenced by the before-mentioned clause, proposed in the Federal Constitutional Convention but rejected, that would have granted the Congress power to promote "agriculture, commerce, trades, and manufactures" (see also the Hamilton quotation, above). Thus a power to regulate one field (commerce) does not imply a substantive power to regulate others (for example, manufactures), nor does it imply a power to *promote* commerce *or manufactures* by spending from the federal treasury, such as promoting and regulating the manufacturing of electricity using atomic energy through a deliberate federal program. The phrase "Commerce among the several States" was meant simply to denote that activity involving the coming and going of merchandise among the states (see Monroe's "Message on Internal Improvements," for example).[38]

Recall that the Supreme Court in *Butler* held that a power to regulate *agriculture* is not among the delegated powers in the Constitution. Indeed, the court emphasized expressly that the *commerce clause* confers no such power.[39] Likewise, commerce is not manufactures. Said the Supreme Court in *Kidd* v. *Pearson* in 1888:

> The language of the grant is, "Congress shall have power to regulate commerce with foreign nations and among the several States," etc. These words are used without any veiled or obscure signification. "As men whose intentions require no concealment generally employ the words which most directly and aptly express the ideas they intend to convey, the enlightened patriots who framed our Constitution, and the people who adopted it, must be understood to have employed words in their natural sense and to have intended what they have said." *Gibbons* v. *Ogden, supra,* at page 188. [9 Wheat. 188]
>
> No distinction is more popular to the common mind, or more clearly expressed in economic and political literature, than that between manufactures and commerce. Manufacture is transformation—the fashioning of raw materials into a change of form for use. The functions of commerce are different. The buying and selling and the transportation incidental thereto constitute commerce; and the regulation of commerce in the constitutional sense embraces the regulation at least of such transportation. The legal definition of the term, as given by this court in *County of Mobile* v. *Kimball,* 102 U.S. 691, 702, is as follows: "Commerce with foreign countries, and among the States, strictly considered, consists in intercourse and traffic, including in these terms navigation, and the transportation and transit of persons and property, as well as the purchase, sale, and exchange of commodities." If it be held that the term includes the regulation of all such manufactures as are intended to be the subject of commercial transactions in the future, it is impossible to deny that it would also include all productive industries that contemplate the same thing. The result would be that Congress would be invested, to the exclusion of the States, with the power to regulate, not only manufactures, but also agriculture, horticulture, stock raising, domestic fisheries, mining,—in short,

every branch of human industry. . . . It is not necessary to enlarge on, but only to suggest the impracticability of such a scheme, when we regard the multitudinous affairs involved, and the almost infinite variety of their minute details.[40]

It is recognized that the *commerce clause* implies some incidental power under the *necessary and proper clause* of Article I, section 8, to regulate manufactures where the *object* of a regulation is an article of manufacture which enters interstate commerce: for example, to prohibit a harmful ingredient; or, in the case of the power industry, to regulate a power plant to ensure that the electricity flowing interstate meets certain specifications of voltage and frequency; or to regulate the interstate shipment of nuclear fuel and waste. But where the object of a supposed regulation of commerce is in fact the manufacturing facility (power plant) itself, such as controls on radioactivity leakage, then such a regulation is not authorized by the present Constitution, as that would imply that Congress has a substantive power to regulate manufactures, and this is not among the enumerated powers delegated by the Constitution. This is not to say that, if the nation is to have nuclear energy, it should not be regulated by a central authority, namely, the federal government. Rather, the question at hand is whether the federal government has at present the constitutional authority to regulate nuclear plants.*

* In *Northern States Power Co.* v. *Minnesota* the issue before the courts was whether the State of Minnesota could, under the Atomic Energy Act, set *more stringent* regulations (limits) respecting radioactivity leakage from nuclear plants within its state than are set by the federal government (AEC). The courts ruled that Minnesota could not, on the ground that the Minnesota regulations would significantly interfere with the other regulatory functions of the AEC under the Atomic Energy Act. (If regulations on leakage were not properly coordinated with other plant health and safety regulations of the AEC, then important operations affecting safety might not receive adequate attention by the plant personnel.) However, in this case the courts were *not* presented with the fundamental issue raised in this chapter of whether the Constitution delegates to the federal government a power to regulate manufacturers—in this case, nuclear power plants. The courts merely assumed that Congress was within its constitutional authority in providing for federal regulation of the "entire spectrum" of nuclear plant operations. The court opinion referred to the *welfare* and *commerce* clauses as the source of this assumed authority, and *Minnesota did not dispute this assumption.* Since, therefore, the constitutionality of the Atomic Energy Act was not challenged and since the Minnesota law would interfere with the federal law, the courts were constrained to uphold the latter, because Article VI of the Constitution declares that constitutional federal laws are supreme (405 U.S. 1035, 447 Fed. 2d. 1143).

For example, in *Hammer* v. *Dagenhart* (1918) the court struck down a federal law which prohibited the interstate shipment of articles made in factories that permitted children to work excessive hours. The court held that the object of the law was not interstate commerce but manufacturing and mining.[41] Incidentally, Congress in 1924 submitted to the states a proposed constitutional amendment that would have granted Congress the power to regulate and prohibit child labor. But the amendment, which was submitted to the state legislatures, never carried.

In 1940, however, the above well-established boundaries of the power to regulate commerce among the states were "overruled" by the Supreme Court in *U.S.* v. *Darby*. The court held that Congress can regulate virtually any activity that "affects" interstate commerce, such as manufacture. Specifically, the court upheld a federal labor law that used the very scheme of regulating labor (prohibitions on interstate shipments, and so on) that the earlier court in *Hammer* v. *Dagenhart* struck down. Said the later court, without any offer of historical argument:

> The power of congress over interstate commerce is not confined to the regulation of commerce among the States. It extends to those activities intrastate ["production of goods"] which so *affect* interstate commerce or the exercise of the power of Congress over it as to make regulation of them appropriate means to the attainment of a legitimate end, the exercise of the granted power of Congress to regulate interstate commerce.[42]

This court opinion, then, is the basis on which the federal government claims a power to regulate the design and operation of civilian nuclear reactors from the standpoint of judging the public safety. For the 1954 Atomic Energy Act, as amended, finds that "The . . . processing and utilization of . . . nuclear [fuel] material *affect* interstate and foreign commerce and must be regulated in the national interest."[43]

This *Darby* opinion or theory is plainly a forced construction of the power to regulate commerce among the states. In 1824 the Supreme Court in *Gibbons v. Ogden* interpreted the *commerce* power plainly enough: "The subject to be regulated is commerce"; and nothing else is to be regulated.[44]

In the Darby case, the court based its opinion solely on a theoretical inference from the 1819 Supreme Court opinion in the case *McCullough v. Maryland:*

> Let the end be legitimate, let it be within the scope of the

> Constitution, and all means *which are appropriate, which* are *plainly adapted to that end, which are not prohibited, but consist with the letter and spirit of the constitution,* are constitutional.[45]

Let us compare the two opinions:

The court opinion in *McCullough* v. *Maryland* asserted in regard to the "incidental or implied powers" of Congress that, if the "end is legitimate," then those *means* which are "appropriate," that is which are "plainly adapted" to the legitimate end and are in the "spirit and letter" of the Constitution, are constitutional. From this opinion, however, the Supreme Court in *Darby* inferred that any means, such as a regulation of manufactures, are automatically appropriate if taken to attain a legitimate end, which is confusing, to say the least, and appears to be a version of the dangerous rule that the "end justifies the means." Here, the criterion for *appropriateness* is switched from those means which are *plainly* adapted and in the *letter and spirit* of the Constitution (the Darby opinion neglects these qualifiers), to any means, just so long as they are taken to attain a "legitimate end," whatever that would mean, no matter how indirect the path. Such sophistry was definitely not to be used as the basis for interpreting the powers of Congress. Said Governor Randolph in the Virginia ratification convention: "No sophistry will be permitted to be used to . . . assume any other power, but what is contained in the constitution, without absolute usurpation." [46] (It is noteworthy that the court in *Butler* cited the same passage from *McCullough* v. *Maryland* to reach a conclusion opposite of *Darby*.)

If sophistry were allowed, the next step, that of construing a power to *promote* manufactures from the commerce clause, could easily be taken, as the Supreme Court did, in effect, shortly after *Darby* in the case of *Wickard* v. *Filburn* in 1942. Without any reference to the intentions of the makers of the Constitution, the court held that "the stimulation of commerce is a use of the regulatory function quite as definitely as prohibitions or restrictions thereon." [47] This theory could easily be relied on to *promote* nuclear energy by spending, without the need for the welfare clause, as the Atomic Energy Commission has asserted.[48] But the Supreme Court in 1824, again in *Gibbon* v. *Ogden*, held that the commerce clause confers "the power to regulate; that is, to prescribe the rule by which commerce is to be regulated." [49] This certainly is no spending power. Recall that the Federal Constitutional Convention rejected a power to *promote* commerce and, indeed, manufactures as well.

It is contended, therefore, that the civilian nuclear program is grounded, not on the intention of those who made and established the Constitution, as to the scope of the delegated powers, or on the early opinions of the Supreme Court, which are closer in time to the making of the Constitution and may therefore be expected to be more consistent with the original intentions, but on the unfounded interpretations of the Constitution in recent judicial opinion.

The question then arises as to whether the latest court opinion is necessarily the true meaning of the Constitution. Certainly, the judgment of the court in a case *is* the law with respect to *that* case,[50] but it is argued that an opinion of the Supreme Court does not bind future courts, as we have learned from the Darby and Gerlach cases, nor does it amount to a constitutional grant of power to Congress. As Blackstone noted in regard to previous court opinions: "*the law,* and the *opinion of the judge,* are not always convertible terms, or one and the same thing; since it sometimes may happen that the judge may *mistake* the law." [51] The Constitution vests in the judiciary only the "judicial Power," which is to "extend" only to deciding cases, that is, specific controversies brought to the court for adjudication.[52] The duty to support the Constitution falls on the judges no more than on any other officer or legislator.[53] As *The Federalist* explained, the courts are to "regulate their decisions" by the Constitution—the "intentions of the people." [54] Clearly, therefore, beyond a particular case decided by the judiciary, the weight of judicial opinion should depend on the force of its reasoning. Having reviewed the relevant opinions, it is maintained that the civilian nuclear program is unconstitutional.

It is well known that the above-mentioned reverses in constitutional interpretation of the *welfare* and *commerce* clauses by the Supreme Court in the 1930s and 1940s were made in order to sustain the measures taken by the federal government to cope with concentrations of economic power and economic depression.[55] But the People are to be the source of new constitutional powers, not the judiciary. We would do well to consider the advice of President George Washington in his Farewell Address:

> If, in the opinion of the people, distribution or modification of the constitutional powers be in any particular wrong, let it be corrected by an amendment in the way which the constitution designates. But let there be no change by usurpation; for though this, in one instance, may be the instrument of good, it is the customary weapon by which free governments

> are destroyed. The precedent must always greatly overbalance in permanent evil any partial or transient benefit which the use can at any time yield.

What would it mean if the federal government's nuclear energy program is unconstitutional? It would mean that the traditional and well-established ways by which the People pursued their safety and happiness, such as requiring liability for accident damages to better ensure responsibility, regulating the manufacturing industry to protect the public health through state or county agency, and controlling what is promoted by their governments by limiting the powers of government, have been discarded without the People's consent. This does not mean that the People will not want to delegate to Congress the power to promote and regulate nuclear energy. It means only that the will of the People has not been determined; and that the People have reserved the right, should Congress want to continue with the civilian nuclear program, to decide the safety issue through delegates close to them, namely, state legislators or state convention delegates.

This concludes the present analysis of the constitutional issue. It is emphasized that it is no more than an outline of this author's argument, presented here only to raise the issue (in perhaps the only effective way). Many very serious considerations have not been addressed, such as: How would a conclusion of the unconstitutionality of the civilian nuclear program be enforced? What could be the role of the Judiciary, and could it be relied upon? If it turns out that the States who might oppose nuclear energy are in a minority but could nevertheless defeat a constitutional amendment, how could their rights be protected if the Congress does not yield to the constitutional process for resolving the nuclear safety issue? These and other important considerations, and the many questions that will surely arise, will have to be taken up in another work, where the matter may be considered in greater depth. To this end, a separate treatise, giving the author's complete constitutional analysis, is being prepared.

Finally, it must be noted, as one may surmise from the preceding analysis, that the constitutional issues being raised here have extremely profound implications, since a great many other large federal programs are founded on the expanded view of the welfare and commerce clauses—programs which, for the most part, have been established since the 1930s, such as social security, aid to housing, aid to education, civilian research and development grants including jet aircraft development, labor standards, superhighways, farm subsidies and regulation, business loan guarantees, civilian space program, foreign aid, airport and airline subsidies, and so forth. Thus, when we question the constitutional method for resolving the nuclear safety issue, the constitutionality of perhaps most of the federal structure as we know it today, aside from the military, will be unavoidably questioned by implication. Again, it is not suggested that, if a particular program is unconstitutional, it is necessarily unwise or contrary to the will of the people, but only that the will of the people through constitutionally empowered representatives has not been determined. Seen in this context of popular control of government policy, which is of growing concern, the nuclear energy issue is just one —though perhaps the most immediately important one—among many policy issues and concerns that demand a recurrence to constitutional principles, in order that the People can best pursue their Safety, Well-Being and Happiness.

Appendix One
The Final Rasmussen Report

THE CRITICAL evaluation of the Rasmussen Report given in chapter 6 pertained to the August 1974 "draft" version of that report. Though it was a draft, the report was nevertheless published by the Atomic Energy Commission and has been widely used by nuclear proponents to justify the safety of nuclear power plants. For that reason, the draft report is critically examined in the present work. The *final* version of the Rasmussen Report has now been issued;[1] and it prompts the following comments.

The final report is as voluminous as the draft version and cannot be evaluated fully without delaying the present work. It appears, however, to suffer from essentially the same shortcomings as its predecessor. This can be seen by examining the final report with respect to power excursion accidents.[2]

Briefly, the final report notes that a power excursion accident at worst could produce fuel melting (in a control rod ejection accident) but that the melting would be confined to a small region of the reactor core (the fuel close to the ejected control rod) and therefore would not be likely to produce a steam explosion that would rupture the reactor. The report admits that the reactor and the containment could be ruptured but assumes that the radioactivity release would be no worse than a slow core meltdown, as in a loss-of-coolant accident without ECCS. Finally, the report argues that the probability of such accidents is extremely small.

The first point to make is that the report analyzes only two limited power excursion possibilities—a control rod ejection (BWR) and a cold water accident (PWR)—and does not consider the more severe possibilities (see pp. 24–30 herein). Recall that the severity of the rod ejection accident depends on the amount of *reactivity* worth of the ejected control rod. Though the report assumes the maximum reactivity worth of any control rod, it assumes the normal vertical positioning of the control rods in the core (called "in-sequence" positions), which minimizes the reactivity worths of individual control rods. This yields a maximum worth of 1.5% (reactivity units), which is still serious since it might produce rapid fuel melting, as the Rasmussen Report notes. (This figure is consistent with this author's previous estimate; see p. 24.) However, *abnormal* control rod positions are possible which can yield much higher reactivity worths—up to 5.5% units[3]—and, presumably, much higher fuel temperatures, more fuel melting, and stronger explosions.

Similarly, the PWR has this possibility of abnormal reactivity worths of control rods, a fact which the report neglects. Though the control rod worths are normally much less in PWRs, since all of the control rods are withdrawn almost totally from the core during reactor operations (which minimizes the reactivity effect of a control rod ejecting the rest of the way out of the core), an abnormally large insertion of one or a few control rods during reactor operation, and hence a greater reactivity worth, is still possible. This possibility is underscored by the fact that one PWR had been operated at power for nineteen days with a control rod fully inserted into the core in violation of the safety specifications.[4] Moreover, at the start of the nineteen-day period the reactor operators were warned that the rod was fully inserted by two separate display-board indicators and by other core data, which had been gathered but not evaluated. Still, these clear indications were carelessly assumed to be faulty and were therefore ignored. Only when the above-mentioned core data were finally evaluated was it realized that the control rod was actually fully inserted, whereupon the reactor was shut down at once.

The faulty control rod was traced to a malfunction in the control rod drive mechanism that allowed the rod to slowly drift into the core. This incident involved many malfunctions and operator errors, and still it happened. In a PWR there is no control rod blocking device to prevent ejection, should the CRDM pipe break.

As for the "cold water accident" in a PWR, the Rasmussen Report concludes that the associated power excursion would present no danger, since the reactivity effect of injecting cold water in the reactor during operation would be so slight that no fuel overheating is even predicted. This conclusion might seem contrary to this author's contention (see p. 26). However, the concern is that the coolant will be not only cold but also free of the boron chemical, which would greatly add to the reactivity potential of the accident. The report neglects this important possibility.

The final report also neglects the severe PEA situations that would produce rapid fuel melting or prompt explosion over a much larger region of the core than in the case of a single control rod ejection. These include the above PWR cold, *unborated* water accident and the BWR autocatalytic steam-bubble collapse possibility (see pp. 25–31). In a rod ejection situation, the fission heating of the excursion is concentrated in the fuel close to the ejected control rod—say, about 5% of the fuel. But in other power excursion accidents all of the fuel would in effect be "running away," as the reactivity changes would effect the whole core, not just a region around a single control rod.

As for the two PEAs addressed in the final Rasmussen Report, no documentation is cited with which one could check the analysis to examine the quality and justification of the various assumptions and mathematical theories that were used. Moreover, the report, again, does not address the question of fuel rod behavior during less severe power excursions, such as the possibility for fuel rod crumbling, the need for experimental verification of predictions, and the possibility for autocatalytic reactivity processes.

As for the *probability* of power excursions, it is important to consider the subjective assumptions inherent in the report's probability analysis of the BWR control rod ejection accident (see pp. 96–97). Explicitly, the report assumes:

1. A .01% chance per reactor year that a control rod drive mechanism (CRDM) pipe will break under pressure.
2. A 1% chance that the control rod movement blocking device is not in place to prevent rod ejection when the CRDM pipe rupture occurs.
3. A .6% chance that the accident will occur during reactor start-up or shutdown operations, when a power excursion could be more severe. (A power excursion initiated at a very low power level allows the reactivity to be greater and the excursion to gain more momentum before the negative reactivity feedback effects set in to stop the excursion.)
4. A 10% chance that the ejected rod will be a high worth rod, though higher worths are possible.

With these factors the report estimates the chances of this severe PEA at two per billion reactor years.

However, the above pipe break factor is based on a probability analysis for *regular* piping, whereas the CRDMs are different—their mode of attachment to the reactor vessel is much inferior (e.g., weaker, partial penetration welds that are not inspectable in service or by radiography, and no vessel wall reinforcement [see Montague PSAR, fig. 4.2-17; Pall. n. 5, p. 128; *Shippingport*, p. 66, see below, p. 227, n. 3; and WAPD-296, p. 84]). Therefore, we should treat the CRDMs independently. Assuming .01% chance of breakage per CRDM per year (upper bound inferable from limited BWR experience) and multiplying by 180 CRDMs per BWR, we get a probability of 1.8% per reactor year. Next, assuming a 2% chance for improper blocking device installation and a 75% chance that a CRDM break will occur (most likely) *during* startups, when such transients impose cyclic pressure and thermal fatigue stresses,[5] and retaining the 10% rod worth factor, we compute a probability value for this severe PEA of 30 per million reactor years (or once every 150 years for 250 BWRs), which is quite high. Indeed, this probability value is even greater than the report's one per million probability for minor, low-consequence accidents (main vol., p. 83).

Another important observation to make regarding the final Rasmussen Report is that the situation of a core meltdown occurring *after* the containment is ruptured is judged much more serious in the final report than it was in the draft report. Recall that one of the shortcomings of the draft report was that it did not analyze for the radioactivity release potential of *prompt* containment failure for many accident possibilities, namely, severe power excursions, power-cooling mismatch accidents, and spontaneous reactor vessel rupture. In these situations, the reactor could rupture or explode while fully pressurized, that is, when the coolant is at its operating pressure and temperature, and rupture the containment promptly. This would mean that the total core meltdown would occur in an open (breached) containment, allowing the escape of the radioactive smoke directly into the atmosphere to maximize the release, since the condensation and trapping of radioactivity within the containment would be minimal. (Incidentally, the final report again neglects this prompt containment failure consideration in its analysis of the BWR control rod ejection accident, by placing the accident in a radioactivity release category of containment rupture occurring after the whole core melts down.)

The draft report, however, does treat one case of containment rupture followed by a core meltdown; but this does not pertain to the above-mentioned accidents. Rather, the assumed accident is a loss-of-coolant-accident in which the containment becomes overpressurized because of an assumed failure to continuously remove the reactor afterheat from the containment atmosphere. (The containment rupture could result in a loss of the emergency coolant for the reactor, which would lead to a core meltdown in an open containment.) Though this would cause a slower core meltdown than, say, a power excursion accident, the radioactivity release to the atmosphere should be large due to the direct escape path. Yet, the draft report assumed a release of only 8% of the highly volatile radioactive iodine, for example.[6]

The final report has reassessed this situation and now predicts 90% iodine release,[7] which seems more plausible. The strontium 90 release is still estimated to be relatively low, only 10% in the final report (up from 3% in the draft report). However, the 3% assumption is based on assuming that 10% of the strontium is released from the reactor in a core meltdown and that 30% of that escapes the containment, whereas the new 10% strontium release estimate is based on assuming a 100% containment escape factor. This reassessment underscores the need to analyze the possibility, indeed, the likely possibility, of prompt containment failure for PEAs and PCMAs, as discussed earlier, especially in regard to the direct release of strontium 90, for these accidents would involve higher fuel meltdown temperatures and, therefore, greater potential for strontium release.

In regard to the measured release of strontium from molten fuel, the final report acknowledges that more than 50% has been observed in some experiments (miniature fuel samples) but argues that 11% is more realistic.[8] However, such is speculation, as discussed earlier (see pp. 83–84, 86). Indeed, in the appendix to the Rasmussen Report which specifically treats fission product release, it is concluded that the uncertainties of fission product properties at very high temperatures and the complexity of fuel meltdown preclude a "highly mechanistic model," meaning that no credible *theoretical* prediction can be made (app. 7, p. VII-7).

As for the land contamination limit for strontium 90, above which agricultural restrictions would be required, the final Rasmussen Report assumes a value which is 150 times greater (less stringent) than that assumed in the WASH-740 report.[9] This is another significant change since the draft report was issued (the draft assumed a 26-fold greater limit relative to WASH-740). Recall that the draft report did not disprove the WASH-740 value, nor did it attempt to derive or justify the value it assumed. The final report, however, includes a derivation of its contamination limit. It is extremely important to scrutinize this derivation, since the implications of the report's value for the contamination limit are very significant in terms of consequences of a severe reactor accident. Crudely, the 150-fold greater limit means that only about 3,300 square miles of land (an area the size of Rhode Island and Delaware) would require agricultural restrictions for the worst conceivable accident, instead of about 500,000 square miles, as indicated by the WASH-740 limit.

Upon examination, it appears that the final report's analysis is founded on at least one faulty assumption and possibly another which could combine to make the calculated contaimation limit much too high, at least by a factor of 375, which could require the use of a limit more in line with the WASH-740 value, indeed, a much more stringent limit.

The Rasmussen Report's calculation of the Sr 90 contamination limit (measured in microcuries of Sr 90 radioactivity per square meter of ground, $\mu Ci/m^2$), denoted by SD (surface deposit), depends on three factors: (1) The total intake of Sr 90 by humans (ingestion) in eating food grown on land contaminated at a level of 1.0 $\mu Ci/m^2$, denoted by *CF*. This is derived on the basis of the soil-grass-cow-milk food chain, corrected for other food grown on the contaminated soil. (2) The radiation dose conversion factor for calculating the number of "millirads" (mr) of radiation dose to the bones per microcurie of Sr 90 ingested, denoted by *DC*. And (3) the maximum acceptable radiation dose, denoted by *RD*, which the report assumes to be 500 millirads per year (mr/yr), versus about 100 mr/yr due to natural radiation.[10] With these factors the report calculated its contamination limit, $SD = RD \div (DC \times CF)$, at 11.7 $\mu Ci/m^2$, versus the .076 $\mu Ci/m^2$ value derived in the WASH-740 report (which includes a safety factor of 2.8);[11] where the Rasmussen Report assumes $RD = 500$ mr/yr, $DC = 42.5$ mr/yr per μCi of Sr 90 ingested, and $CF = 1.005$ μCi of Sr 90 ingested per $\mu Ci/m^2$ on the ground. We shall now critically examine these assumptions, particularly *DC* and *RD*, the dose conversion factor and the allowable radiation dose, respectively.

Firstly, the Rasmussen Report's dose conversion factor, 42.5 mr/yr/μCi, is not consistent with the value derivable from the Federal Radiation Council's data on dose conversion, that value being 792 mr/yr per μCi of Sr 90 ingested.[12] This means that for a given amount of Sr 90 ingested the bones would actually receive an 18.6 times greater rate of radiation dose than the report's dose conversion factor would predict ($792 \div 42.5$). The FRC-based value can be checked easily, since it is straightforward to derive; whereas the Rasmussen Report's value is based on "elaborate computer models" which cannot be readily evaluated.

For those wishing to check this important discrepancy, the following data were used to calculate the dose conversion factor, *DC*:

Average beta ray energy per strontium disintegration, including the energy of the short-lived Yttrium daughter:

> .98 MEV, based on assuming the average-to-maximum (end-point) ratio for the beta particle energy of .35 (R. D. Evans, *The Atomic Nucleus*, p. 540).

$T_{1/2}$ = biological half-life of strontium 90: six years (see WASH-740, p. 35).

Mass of bones: 7 kg (WASH-740, p. 36).
1.0 MEV $= 1.6 \times 10^{-6}$ erg of energy.
1.0 mr $= .001 \times 100$ erg/gm of tissue (definition of millirad).
1.0 μCi $= 3.7 \times 10^4$ disintegrations of Sr 90 per second (definition of μCi).
$Q = 200 \times 10^{-6}$ μCi of Sr 90 ingested per day results in 500 mr/yr radiation dose to the bones (*Radiological Health Data and Reports* 10, no. 7 [July 1969]: 301.) This conversion rule implies that 30.3% of the ingested Sr 90 settles in the bones, which is consistent with the WASH-740 value of 22% for the inhalation-lung-blood-bone path (WASH-740, p. 36).

The dose conversion formulas are:

$dS/dt = f_i \times Q - \lambda S = 0$ (equilibrium); $\lambda = \dfrac{\text{Log } 2}{T_{1/2}}$; and dose rate $= G \times S = \dfrac{G \times f_i \times Q \times T_{1/2}}{\text{Log } 2}$, mr/yr; where f_i is the fraction of Sr 90 ingested that settles in the bone; S is the amount of Sr 90 in bones; and G is the dose conversion coefficient, mr/yr per μCi of Sr 90 located in the bones.

$$G = \frac{3.7 \times 10^4 \times 365 \times 24 \times 60 \times 60 \times .98 \times 1.6 \times 10^{-6}}{.001 \times 100 \times 7{,}000}$$

$= 2{,}614$ mr/yr per μCi of Sr 90 in bones.

$$f_i = \frac{.693 \times 500}{6 \times 2{,}614 \times (200 \times 10^{-6} \times 365)} = .303.$$

Result: $DC = f_i \times G = 792$ mr/yr per μCi of Sr 90 ingested, to be compared with the report's value of 42.5 mr/yr/μCi.

Secondly, the maximum acceptable dose of 500 mr/yr is twenty times higher than the maximum safe level of radiation exposure now being proposed by the U.S. Environmental Protection Agency, namely, 25 mr/yr, which is another difference factor[13] which we might want to apply. Combining these differences, we need to reduce the Rasmussen Report's contamination limit by 370 times, which brings it in line with the WASH-740 value; indeed, the corrected value is more stringent than WASH-740, namely, .03 μCi/m^2 versus .076 μCi/m^2, which are to be compared with the report's value of 11.7 μCi/m^2. Alternately, if we use a 170 mr/yr limit, which is the generally accepted limit for the general population,[14] and WASH-740's safety factor of 2.8 along with the dose conversion factor correction, we would have to reduce the report's value by 150, which returns us to the WASH-740 limit.

The food chain factor, *CF*, remains to be scrutinized, but this cannot readily be done. It is noted only that the theoretical model of the food chain used in the Rasmussen Report neglects, without a numerical error analysis, the redeposit of strontium in the soil by cow excretion, which would slow the rate of depletion of Sr 90 from the soil relative to what the model predicts, and consequently would lead to a higher radiation dose over time, requiring a further reduction of the contamination limit. This analysis also neglects other sources of radiation exposure following the passage of a radioactive cloud, such as inhaling radioactivity and eating foods more directly contaminated, which would reduce the tolerance for additional dose from foods contaminated by uptake of Sr 90 from the soil.

We could conclude, therefore, that the Rasmussen Report's maximum acceptable contamination level for Sr 90 is grossly high, by about 150 times or possibly more, which means that the WASH-740 damage criterion for Sr 90 land contamination, which this author has used to estimate the consequences of reactor accidents in terms of land area requiring agricultural restrictions, still remains valid. It is urged that health and pollution authorities outside the nuclear energy establishment scrutinize this particular volume of the Rasmussen Report (app. 6) to identfy and evaluate the adequacy or wisdom of the various assumptions and conclusions underlying the estimates of public harm and damage. Toward this end, it is recommended that the report be expanded to represent its analysis of accident consequences in a form which follows identically the systematic analysis of the WASH-740 report—that is, a parallel analysis, so that one can identify readily and precisely the specific differences between the two analyses. This parallel analysis should also address all of the questions and uncertainties noted in the WASH-740 report.

Appendix Two
Summary Description of Fourteen Accidents and Near Accidents in Nuclear Reactors

PRESENTLY, the U.S. has about forty-five large-size light water reactors (LWRs)—PWRs and BWRs—and fifty more under construction. There have been several near-accident incidents with the LWRs. The U.S. has had three LMFBRs: Experimental Breeder Reactors I and II (EBR-I and EBR-II) and the Fermi demonstration LMFBR power plant. Both EBR-I and Fermi suffered a fuel meltdown accident, and there have been three accidents in other reactors. Of the various accidents and near-accidents that have occurred, the fourteen cases described below are probably the most significant; though this does not mean that other incidents were not serious.

1. EBR-I Core Meltdown

The first experimental breeder reactor (EBR-I) suffered a core meltdown in November 1955. The reactor came within a half-second of exploding before being brought under control by the actuation of a back-up, fast-acting, reactor shutdown (SCRAM) system. However, the incident posed no real threat to the public safety since the EBR-I core was very small, about ¼ cubic foot, compared to the 123-cubit-foot cores of large LMFBRs; the power level was small, specifically, 1.4 megawatts of heat output, compared to the 3,300 megawatts of today's large reactors; and the reactor was located in an Idaho desert and used no plutonium.

The incident took place when the reactor power was being increased to about one half of its design full-power level in a special experiment, which was to investigate a troubling instability phenomenon that had occurred and had not been anticipated in the design. During this process, the power level suddenly began to rise on its own accord. This was later determined to have been caused by the following phenomenon: The heat-up of the fuel rods caused them to bow inward due to thermal expansion effects, and this effectively compacted the fuel and thereby increased the *reactivity* of the system, which caused the power level to rise. The higher power level then made the rods bow even more, thereby accelerating the process beyond the control of the reactor operator, who hesitated and then SCRAMmed the reactor in an attempt to reduce the reactivity and shut down the reactor—that is, stop the fissioning. (The SCRAM consisted of rapidly inserting the control rods into the reactor core, which normally would have shut it down.) However, though the SCRAM system functioned, this action did not reverse the power rise. Apparently, fuel melting by overheating had occurred by then (fuel melting, as well as inward fuel-rod bowing, can increase reactivity in an LMFBR, as both processes tend to compact fuel, which tends to raise reactivity); or increased fuel bowing, other subtle effects, or some combination occurred, which increased the reactivity enough to overwhelm the reactivity reduction of the SCRAM.* (If the reactivity should rise beyond a threshold called "prompt critical," a strong and rapid power rise called a nuclear excursion would be triggered, which could produce an explosion, depending on circumstances. In such an explosive nuclear excursion, the fuel would be heated greatly and would thus be vaporized at explosive pressures.)

Fortunately, EBR-I was equipped with a back-up SCRAM capability (consisting of dropping the "blanket" away from the core, to strongly reduce the reactivity), although the reactor operator failed to use it, which allowed the reactivity to approach dangerously close to prompt critical. A scientist observing the situation quickly reached over and pushed the back-up SCRAM button that finally shut down the reactor. If he had acted a half-second later, a nuclear excursion would have occurred, as the reactivity was rising quickly. Such an excursion would have produced an explosion, with a force of about 1 pound of TNT explosives. Allowing for uncertainty, an explosion of about 100 pounds TNT is probably an upper bound of the maximum pos-

* Subsequent LMFBRs are designed with more control rod SCRAM reactivity (negative).

sible that could have occurred in EBR-I, which is very much less than the 20,000 pound TNT-equivalent conceivable explosion potential of large LMFBRs.

However, because of the fuel overheating, the core melted down and was therefore destroyed (see fig. 14). When the back-up SCRAM occurred, the power level dropped to zero, and the core froze in a heap of debris. Because there was no explosion, the reactor vessel and associated coolant piping system remained intact (sealed), so that the radioactivity was confined to the reactor system.

The rod-bowing problem has been eliminated (at least for Fermi and EBR-II) by design features (fuel rod restraints) that will hopefully avoid this type of autocatalysis in the future, but whether this or some other effect will occur unexpectedly in future LMFBRS remains to be seen (see pp. 146–49). Also, the cause and progression of the EBR-I power excursion is far from being understood, as discussed in the recent article by G. S. Lellouche.[1]

2. Fermi Reactor Partial Fuel Meltdown

On October 5, 1966, the Fermi LMFBR, which is located twenty miles from Detroit, Mich., suffered a partial fuel meltdown when several pieces of sheet metal broke off the bottom of the reactor vessel and were swept up by the coolant flow to become lodged under the core. The sheets then choked off coolant flow into two of the 105 fuel elements in the core. The affected fuel elements overheated and melted, as the reactor was being operated at a substantial power level (15% of full power).

The danger of a fuel meltdown in an LMFBR is that a severe nuclear explosion is potentially possible following rapid core compaction, which could occur by gravity collapse of a molten or weakened core. Such an explosion could be as high as 1,000 pound TNT-equivalent or higher for the Fermi design and could rupture the outer shell around the reactor, the containment, releasing hazardous radioactivity into the atmosphere. A University of Michigan study produced for the Fermi designers concluded that, if this happened, as many as 115,000 people could be killed in nearby cities. The WASH-740 estimates of land contamination and its effects would apply, since the Fermi was of the size assumed in that analysis.* (Fermi started with no plutonium.)

* Fermi has been permanently shut down ("decommissioned").

Figure 14. EBR-I Core after the meltdown incident.

The Fermi incident was aggravated by several human errors. First, when the fuel overheating was first observed by temperature dials, the operators did not heed the trouble indication. Then, when the fuel began to melt, the reactivity decreased, as the initial melting did not produce a net core compaction. The power level began to drop as a result, which was an abnormal behavior of the core, since the operator did not cause the drop—that is, the power dropped spontaneously. Instead of SCRAMming the reactor (inserting control rods),the operator *raised* the control rods to increase reactivity and maintain the high power level, which aggravated the fuel melting. The melting was occurring without the operators realizing it. However, the control rods were observed to be in an abnormal position, indicating an anomalous reactivity effect. Radioactivity leakage from the two overheated fuel elements then set off a high-radiation alarm. Still, the operators maintained a high power level. Eleven minutes after the radiation alarm, the anomalous reactivity reversed its direction and began to rise slightly, indicating that the fuel-melting effects were beginning to cause a net core compaction or that some other subtle effect was occurring (see fig. 5 of the article by R. L. Scott, n. 2, below). With the situation thus becoming more uncertain, especially with respect to reactivity, the reactor was finally SCRAMmed.

Had the reactor not been SCRAMmed, or if the SCRAM system had failed to function, a runaway power-level situation could conceivably have ensued, as in the EBR-I case, to produce a core-wide meltdown and compaction. This in turn could have produced a severe nuclear excursion and explosion. At that particular time, there was relatively little radioactivity, compared to the amount assumed in the WASH-740 analysis—for example, about 1% of the strontium 90 quantity assumed in WASH-740—because the reactor had been operated very little prior to the accident. Therefore, the consequences of an explosion may not have been too severe. However, the accident could just as easily have occurred after the reactor had operated for a long time and when the reactor was at full power. Under these circumstances, the radioactivity hazard would have been great; and the potential for fuel melting would have been worse.

The sheet-metal pieces that caused the incident were installed at the bottom of the reactor under the core as a hurried, last-minute, design change. The sheets were very thin and were held down by the heads of several screws. Evidently, the coolant flow lifted the sheets, which easily tore away from the screws to be swept up by the coolant.[2]

3. Shippingport (PWR) Steam Generator Drop

In 1964 the Shippingport PWR reactor near Pittsburgh, Pa., was modified to increase the power output from 230 megawatts to 500 megawatts of heat. Among other changes, the original steam generators were replaced with heavier models. The steam generators are massive, 60-ton, heat-transferring equipment that use the heat of the reactor coolant to make steam; the coolant is then pumped back to the reactor via connected piping. Since the coolant piping could not carry the weight load of the heavy steam generators, they were suspended from "support hangers" attached to an overhead. However, the hangers were faulty and broke when the new, heavier steam generators were being filled with coolant in preparation for reactor operation at the higher power levels. An alert workman in the steam generator compartment sounded the alarm, which prevented the total fall of the steam generators. Had the hangers failed during power operations, the connected main coolant pipes would have been ripped off, and the worst loss-of-coolant accident would have ensued. (Due to high radiation levels in the steam generator compartments when the reactor is producing power, there would have been no workmen present to notice the hanger failures.) These hangers were faulty even for the prior service but luckily did not break then, and their adequacy for reuse with the heavier steam generators had not been checked.

The danger of a loss-of-coolant accident is, of course, that the emergency core coolant system (ECCS) might not function, or might function ineffectively; the core would then heat up and melt down, causing serious public harm by a major release of radioactivity.[3]

4. Hanford Reactor Failure to SCRAM

On September 30, 1970, trouble developed in one fuel element of the Hanford "N" reactor—a loss of coolant flow. Detection generated an automatic reactor SCRAM signal, but all eighty-seven control rods failed to insert into the reactor. Fortunately, it was equipped with a back-up SCRAM system of a totally different design, which was then called upon and functioned properly to shut down the reactor safely. (The Hanford reactors are used to produce atom-bomb material and hence are fundamentally different than commercial reactors. None of the commercial PWRs and BWRs has a back-up SCRAM system.)

The failure to SCRAM was caused by a short circuit in the electrical circuitry of the SCRAM system. Normally, such a short circuit would render only the affected circuit component inoperable, leaving other back-up or redundant components in the circuitry operable, and these redundant components would have been sufficient to actuate the SCRAM system. That is, the SCRAM circuit was designed to be fail-safe. But the molten metal generated by the electrical short flowed, then froze, in such a fashion as to form a new viable circuit that had the effect of blocking the SCRAM function. Specifically, two unrelated electrical wires were connected by the frozen metal, which allowed one wire to feed electricity to the control rod drive motors to keep them from SCRAMming. The SCRAM circuit, however, was designed to *deenergize* the motors by opening switches that cut off the electrical supply to the drive motors and thus allow the control rods to drop into the reactor by gravity. But this design function was negated by the fluke short circuit. This shows how the craziest things can happen to render safety systems inoperable.[4]

5. BWR SCRAM System Failure

A SCRAM system for a BWR was found totally inoperative in a routine test. This failure condition had existed for about two weeks before being detected.

The SCRAM system is the most important reactor safety system. However, the BWR has no back-up system, which is justified on the grounds that the system contains several back-up electrical switches, either one of which would SCRAM the reactor upon opening (the SCRAM instrumentation acts to open all the switches upon a trouble signal). However, all of the switches were manufactured improperly—specifically, one error in the manufacturing process caused the coatings on the switches to become sticky with the passage of time. When the SCRAM system was tested in a periodic test, *all* switches were found to remain closed (stuck).

If any number of reactor malfunction incidents which require a SCRAM had occurred during the period of the inoperative SCRAM system, a disastrous reactor accident would have resulted. For example, on May 27, 1971, the Millstone BWR at Waterford, Conn., suffered a malfunction of the steam valve which caused the reactivity, and hence the power level, to rise beyond the full or rated power level. The safety instruments detected the power rise and SCRAMmed the reactor. Had the SCRAM system failed, a runaway power level could have ensued. The potential for such a runaway power transient has not been scientifically bounded—except that it could potentially result in rapid core melting and

and nuclear runaway explosion, and finally in a major public disaster. Such is the significance of a SCRAM system failure.

Though the AEC never traced the origin of the manufacturing error, they suspect that the faulty switches were the result of the manufacturing plant being moved to a different location. The new personnel were not as experienced as the old, and this could have accounted for the error in manufacture. Prior to the move, the company had had a record of thirty years without significant failure of its switches—which is why they were selected for use in reactors. The incident thus shows how reliable equipment can suddenly become unreliable. General Electric Co., the makers of the BWRs, testified before the incident occurred, or was made known, that the probability that the BWR SCRAM system would fail is 10^{-10}, or one failure in every ten billion attempts. Yet, the above incident indicates a 10^{-4} failure probability, or one in every 10,000 attempts. With a projected 500 LWRs, a 10^{-4} failure probability is very serious, for if we assume two incidents requiring SCRAM per year per reactor (too generous to the industry), 500 reactors would mean at least one major accident every ten years. The AEC in 1973 issued a regulation requiring back-up SCRAM systems (though not in a complete sense) for reactors whose license applications are submitted after 1977, but this will leave 200 reactors without back-up systems. Moreover, it is possible that the failure record will get worse with the passage of time, as the industry becomes lax or overconfident—that is, complex multiple failures can occur, as is well founded in human experience.[5]

6. B-52 SAC Bomber Crash Close to a BWR

In January 1971 a B-52 bomber was flying a routine practice flight over a small BWR located near Charlevoix, Mich., when it crashed in Lake Michigan about two miles from the reactor (Traverse Bay). An eyewitness said the plane was heading directly in line with the reactor when it crashed, skipping off the surface of the water, and raising a fireball 200 to 600 feet into the air. If the plane had crashed into the reactor, there would have been a major public disaster, since a loss-of-coolant would have occurred at the very least. Radioactivity release would have been promoted by the punctured reactor container and the burning petrofuel.

In 1967 the bombers and fighter-bombers were routinely (300 flights a month) flying low-level (1,000 foot) flights directly over the plant, despite U.S. Air Force instructions to the pilots to stay clear of the plant, and the practice continued until the crash in 1971. It has been speculated by a Grumman aerospace official that the plane may have flown into radioactive gases normally discharged by the reactor plant's effluent stack. The radioactivity could have interfered with the plane's electronic guidance systems (a process called latching), which might then have caused the pilot to misjudge its altitude. No report of the cause of the crash has been made public.[6]

7. Vermont Yankee BWR Criticality Incident

The Vermont Yankee reactor accidentally went "critical" during a maintenance operation when the reactor was shut down, and almost suffered a power excursion with both the reactor vessel closure head and the containment dome off. The reactor vessel and containment are required to be sealed whenever the reactor is made critical, so that should the reactor suffer a design basis, power excursion accident, the vessel and containment would prevent the serious release of radioactivity into the atmosphere (according to prediction). Fortunately, the SCRAM system functioned to shut down the reactor safely.

The incident occurred when the reactor operator was withdrawing a single control rod to test its SCRAM ability (each control rod is so tested during maintenance outages). The reactor is so designed that a total withdrawal of one control rod will not make the reactor critical, that is, it will not produce fissioning. The rest of the control rods, of course, are to be fully inserted into the core when the reactor and containment are open. Furthermore, an electrical "interlock" safety device is installed which prevents the reactor operator from withdrawing (raising) a control rod if any other rods are not fully inserted in the core. However, to save time, the operators violated safety procedures by installing a "jumper cable" that negated the interlock in order to permit another, semirelated operation. Then two more human errors occurred: the reactor operator in one work shift pulled one control rod and inadvertently left it fully out; the operator in the next shift failed to notice the condition on his instrument board, nor was he informed of the status of the withdrawn control rod. Also, the maintenance crews failed to remove the jumper cable, and when the reactor operator proceeded to raise an adjacent control rod, the reactor went critical. Had the cable been removed, the second rod could not have been raised.

The circumstances were such that, even if the SCRAM had failed to occur, there may not have been a serious accident involving the public, though the reactor could have been damaged or the reactor building heavily contaminated. Given the many human errors involved, however, the circumstances could have been worse. For example, a control rod dropout accident could have occurred, if the control rod being withdrawn had accidentally disengaged from its drive shaft and fallen out of the core. After all, the purpose of the routine control rod testing that was being carried out, and which brought about the incident, was to determine whether the control rod was properly engaged with its shaft. Furthermore, such an incident with a SCRAM failure has never been calculated for the course it could take.[7]

8. Loss-of-Coolant Incident in a BWR

A malfunction of a steam valve led to a sequence of reactor system valves and pumps turning on and off, during which a divergent (runaway) oscillation of the water level in the reactor vessel occurred. At one point, the water dropped below the core (the fuel), leaving the core without coolant. Fortunately, the reactor control rods were inserted early in the transient, which stopped the heat production (fissioning) well before the water level dropped (ten minutes); also, there was little afterheat, because the reactor had been shut down for seven months and had been operating only twenty-eight hours when the incident occurred. Had the reactor been operating for a week or longer, there would have been enough intense short-lived radioactivity accumulated in the core that the afterheat would have presented a hazard. At ten minutes after fission shutdown, the afterheat could have been 3% of the full power level. Without coolant in the core, the fuel could then have heated up and melted in four minutes, with the core crumbling well before that (two minutes). Crumbling could prevent adequate cooling even if the water level were restored, and this would then lead to core meltdown, which can end in a major public disaster. However, there is insufficient information in the AEC's report of the incident to determine how long the core was uncovered by water, in order to then determine whether the core could have had time enough to heat up to the danger point with the after heat. Hence, we must presume that the incident could have been disastrous. (This point must be further explored.)

The water level dropped partly because some steam valves opened accidentally (caused in part by a human error) and partly because pressure transients in the system occurred that caused all of the pumps that inject water into the reactor to quit. The result was a heavy discharge of steam (boiled coolant) from the reactor without any make-up water coolant. The water level therefore dropped rapidly to below the core before the operators could react to close the steam valves and restart coolant injection pumps. Had there been substantial afterheat present, the drop in the water level would have been even more drastic, since the heat would have boiled away the water even faster. Also, the afterheat would have been more intense, since the core would have uncovered sooner after the SCRAM.[8]

9. BWR Containment Pressurization

Though this incident constituted no serious public hazard, it does illustrate well how a train of human errors can develop after a reactor departs from normal conditions, which might well have occurred under more hazardous circumstances. A rough sketch of events is as follows:

The steam valve began to malfunction and then closed. Fortunately, the reactor SCRAMmed automatically. The power dropped to the afterheat level, reducing the size of the steam bubbles in the core. This caused the water level in the reactor to drop, which caused the feedwater pumps to increase coolant flow into the reactor to avoid uncovering the core. As the water level rose, the operator noticed that the level indicator was reading a low level. Actually, however, the indicator was stuck and giving a false low-water-level reading. The operator reacted by manually increasing the feedwater flow still further, so that the water then filled the reactor and spilled over into the steam line. The feedwater-flow error was uncovered and corrected; but then the pressure began to rise, and two safety systems designed to cope with the problem and cool down the reactor were found inoperative. The operator then reduced pressure by opening a relief valve momentarily. At this point, water hammer occurred, produced by the water spill-over into the steam line, and this popped safety valves (pressure relief valves), which stuck open due to a design error. The relief valves then discharged reactor steam to the reactor containment atmosphere, which began to pressurize the containment. The loss of coolant through the stuck relief valves should have caused the ECCS to activate to inject replacement coolant; but one system was found inoperative, and the operators blocked the operation of the other system on the assumption that the loss-of-coolant problem was minor. However, they did not know the cause (stuck valve) and could not make a sound judgment (it could have been a leaky coolant pipe about to completely rupture). Meanwhile, the pressure in the containment rose beyond the range of the pressure gauge (5 psig). The containment is equipped with water sprays to quench the steam pressure whenever two psig pressure is exceeded, but the operators blocked this safety action because that would have cold-shocked some equipment and thereby damaged it. They did not, however, have sufficient knowledge of the events to justify their action. The containment reached 20 psig compared to 60 psig design pressure before the plant was finally brought under control.

The lesson to learn from this incident is that human errors can quickly develop and be compounded to lead to unanticipated situations. This could just as easily occur in a really hazardous situation to make matters worse.[9]

10. Browns Ferry BWR Cable Fire

The Browns Ferry reactor complex in Alabama consists of two large BWRs that share the same control room. The electrical cables, which send control signals to the reactors and their safety systems, and operational data to the control room from these systems, converge in a cable room under the control room. Workmen who were routinely plugging air leaks in the outside wall of the cable room used a candle flame to test for leaks. The flame ignited a plastic sealing material, and the fire spread into the cable room through the air leak, igniting the cables. Most of the cables were destroyed, rendering most of the safety systems inoperable. The reactors were both operating at full power at the time (2,200 megawatts of electricity total, which serves about two million people). The operators lost control of one reactor for a time but were able to regain bare minimum control just in time to avert uncovering the reactor core of water, thereby preventing a possible core-meltdown accident and a possible major public disaster. After the fire was put out, the operators struggled for sixteen hours to bring the reactors under control. It appears that it was only a matter of luck that the fire did not knock out control of the few safety devices that remained in operation; as it was, the crew had to resort to considerable makeshift.

It is of interest that a cable fire occurred in an earlier plant, which rendered certain redundant (back-up) safety systems inoperative. The lesson that was learned was that the cables of back-up safety systems should not be located so close to the cables of the primary systems that a single error (a fire) could knock out both systems at once. This lesson was evidently not applied to the Browns Ferry complex or, presumably, to other reactors, since they all presumably have a single cable room in which all cables converge. Indeed, prior to the fire at Browns Ferry, the propriety of using a candle to check for air leaks was discussed at the plant management meetings, including the fact that a previous fire had occurred but was put out in time; yet, no corrective action was taken.

The draft Rasmussen Report, issued prior to the incident, noted that a fire once occurred in a cable tray and acknowledge the need for the "routing and separating of the safety system cables," but the report's probability analysis apparently assumed that a fire will affect essentially only one cable tray, not an entire room of trays. To modify existing plants and those under construction to eliminate this weak point would cost an estimated $10 billion, according to D. Comey. No decision has yet been reached as to whether such modification will be required.

Accounts of the Browns Ferry fire accident shows again how trains of common human error and malfunctions can pile up. A rough sketch is as follows:

The fire started while both reactors were at full power. There was a fifteen-minute delay in sounding the fire alarm to alert the reactor operators and fire fighters, during which time the operators could have initiated safety actions. The delay was caused in part by faulty equipment and procedures.

After the fire alarm, the reactor operators waited five more minutes before shutting down the reactors, which they accomplished by inserting control rods, despite the facts that the control board was behaving erratically (flashing lights and alarms) and that the emergency core cooling systems were turning on and off.

The afterheat in the core still had to be removed by cooling the reactor; otherwise, the reactor coolant would overheat and generate excessive pressure. However, reactor shutdown cooling systems were lost. The reactor pressure relief valves were then activated, but this made makeup coolant necessary and the emergency cooling systems then were lost. A cooling system was jury-rigged using a low-pressure steam condensate pump that was never intended for such use, but this required opening more steam relief valves further to reduce the pressure, so that the low-pressure pumps could inject water into the reactor. This measure involved the risk of uncovering the core due to the more rapid coolant loss through the added relief valves until the pumps could restore the reactor water level. The water level fell to within 48 inches of uncovering the core, compared to 200 inches of normal water cover. Fortunately, the fire had not destroyed all of the water-level and pressure indicators (most were lost); if it had, the operators would not have known what kind of action to take.

Other malfunctions and problems occurred as well, contributing to the overall delay in bringing the reactors under control.

The NRC's "Evaluation of Plant Safety during the Browns Ferry Fire" concludes that if alternate action had been taken by the plant operators in response to the fire, there would have been a greater margin of safety during the efforts to bring the reactors under control. But this evaluation was made after the fact, when the NRC had had time to study the safety capability of the plant relative to an unthought-of accident. During an actual accident, one does not have time to fully assess the safety capabilities. One must act; and such action can always be the wrong action.[10]

11. Cracks in the Emergency Core Cooling System Piping at a BWR

In January 1975, during a special inspection, cracks were found in the high-pressure piping of the emergency core cooling system (ECCS) of the Dresden No. 2 BWR reactor in Illinois. The reactor had been shut down for the inspection. The observed cracks penetrated the full thickness of the piping, as evidenced by water coolant leakage that was found when the pipe insulation was removed. The cause of the cracking has not yet been determined. Since the ECCS is a vital safety system, the NRC ordered the shutdown of twenty-two other BWRs for inspection of their ECCS piping (no cracks were found, however).

The hazard created by the pipe cracks is shown as follows: Should an ECCS pipe suddenly break off completely at full reactor power due to an accelerated crack growth, the hot, high-pressure reactor coolant would blow out of the ruptured pipe, exploding into steam. This would be a loss-of-coolant accident. In any such loss-of-coolant incident, the ECCS is designed to reflood the core before the fuel temperatures become excessive, though it is normally assumed that the coolant recirculation pipes would rupture, not an ECCS pipe. The significance of an ECCS pipe rupture is that it not only causes a LOCA but knocks out an ECCS system at the same time, since a ruptured ECCS pipe cannot carry emergency coolant into the reactor. (However, there are two or three back-up or "redundant" ECCS systems.) Furthermore, if all ECCS piping had cracks and if a LOCA occurred—say, by the rupture of a main reactor coolant recirculation pipe—the reactor vessel movements reacting to the 800,000-pound jet thrust of the expelling coolant might impose enough extra stresses in the cracked ECCS piping to cause them to rupture in response, thus rendering all of the ECCS systems ineffective (This possibility has never been investigated theoretically by the nuclear community in the open literature; also, there may be other sources of pipe stresses in a LOCA.) This could be why the order was given to shut down the twenty-three BWRs for inspection. Moreover, the cracking problem might have been common to the coolant piping, thus increasing the chance for such a scenario.

The discovery of the ECCS pipe cracks in Dresden No. 2 was accidental; that is, it was not the result of any periodic inspection program. Large cracks had occurred in other unrelated piping, which leaked noticeable quantities of water, and similar cracks were then found in many other BWRs in the same piping. This triggered an investigation of other piping, including the ECCS piping, which lead to the discovery of the ECCS pipe cracks.[11]

12. Loss-of-Coolant Accident in a Gas-Cooled Reactor (Switzerland)

A small, 30-megawatt reactor built inside of a mountain at Lucens, Vaud, Switzerland, suffered a loss-of-coolant accident on January 21, 1969 in which the core suffered limited damage. A heavy amount of radioactivity was released into the reactor cavern, which was quickly sealed. The latest available report was issued in the summer of 1969; and the status of the reactor is not known.[12]

13. SL-1 Reactor Runaway Explosion

In 1961 a small experimental reactor underwent an accidental power excursion which caused the reactor to explode, killing by blast and radiation, three maintenance workers standing on top of the reactor. The accident occurred partly because the workers carelessly or unknowingly raised a control rod too far out of the core during a maintenance operation. Present LWRs (PWRs and BWRs) are considerably less hazardous than the SL-1 reactor in such situations, due to certain design differences, but nevertheless, today's large LWRs have a theoretical potential for explosive nuclear runaway, which can cause major public disaster. Such accidents can be caused by carelessness, such as occurred in the accidental criticality of the Vermont Yankee BWR. No public harm was caused by the SL-1 accident, due to the relatively small amount of radioactivity, the absence of afterheat, and the reactor's location (Idaho desert).[13]

14. Windscale Reactor Meltdown (England)

Since this reactor is somewhat similar to the HTGR (pp. 152–53), the incident should be mentioned. Graphite core material caught fire and released large quantities of volatile iodine radioactivity, which forced a ban on dairying for sixty days over 200 square miles of land.[14]

References and Notes

Chapter One

1. R. E. Webb, "The Explosion Hazard of the Liquid Metal Cooled, Fast Breeder Reactor (LMFBR)" (1973; limited circulation), discussed in USAEC WASH-1535, *Proposed Final Environmental Statement* (PFES), Liquid Metal Fast Breeder Reactor Program (Dec. 1974), 6: 38-256, 257, 38-276–91; and 2: 4.2-155; see also *Environment* 16, no. 6 (July/Aug. 1974): 6, 10; and 17, no. 4 (June 1975): 8–11. R. E. Webb, statement on the LMFBR Demonstration Plant, U.S. Congress, *Hearings before the Joint Committee on Atomic Energy,* 92d Cong., 2d sess. (Sept. 1972), in "The LMFBR Demonstration Plant," app. 5 (hereafter cited as *Hearings*).
2. WASH-1400, *Reactor Safety Study,* USAEC (Aug. 1974; draft). H. W. Lewis et al., "Report to the American Physical Society by the Study Group on Light-Water Reactor Safety," *Reviews of Modern Physics* 47 (July 1975) (hereafter cited as APS Rpt).
3. WASH-1535, vol. 2, sec. 4; vol. 3, sec. 6A; vol. 4, sec. 7; and vol. 6, sec. 38.
4. WASH-740 (1957), p. 2.

Chapter Two

1. BMI-1910, D. L. Morrison et al., "Core Meltdown Evaluation" (Battelle Memorial Institute, July 1971), app. D.
2. Ibid.
3. Ibid., app. C.

Chapter Three

1. WASH-1400 (Rasmussen Report), app. 1, pp. 141–42, 195–96, and app. 8, subapp. C.
2. See generally WASH-1400. See also T. J. Thompson and J. G. Beckerley, *The Technology of Reactor Safety* (MIT Press, 1973), 2, chap. 21: 785–87.
3. WASH-1400, app. 5, pp. 109–72; and Thompson and Beckerley, *Technology,* vol. 2, chap. 19.
4. MI-1910, app. C.
5. PTR-738, G. O. Bright et al., "A Review of the Generalized Reactivity Accident for Water-Cooled and -Moderated, UO_2 Fuelled Power Reactors," internal report (Phillips Petroleum Co., Atomic Energy Division, National Reactor Testing Station, USAEC; undated, though issued before Feb. 1965, according to library marking).
6. WASH-1400, app. 8, subapp. C.
7. Herbert J. C. Kouts, "AEC's Nuclear Safety Research Objectives, Plans, and Schedules," *Nuclear Safety* 15, no. 2 (Mar./Apr. 1974): 128–29. See also article by C. K. Leeper in *Physics Today* (Aug. 1973), pp. 31–35.
8. APS Rpt.; D. F. Ford and H. W. Kendall, *An Assessment of the Emergency Core Cooling Systems Rulemaking Hearing* (Ballinger, 1975). H. W. Kendall, "Nuclear Power Risks: A Review of the Report of the American Physical Society's Study Group on Light Water Reactor Safety," Union of Concerned Scientists (June 18, 1975).
9. Thompson and Beckerley, *Technology* 1: 176. See also PTR-738.
10. Thompson and Beckerley, *Technology,* 2: 805, 808.
11. Inferred from AEC doc. IDO-17028, J. E. Grund, ed., "Experimental Results of Potentially Destructive Reactivity Additions to an Oxide Core" (Dec. 1964), pp. 18–46, 85.
12. Bailly 1 BWR, Preliminary Safety Analysis Report, "Maximum Rod Worth versus Moderator Density," AEC docket no. 50-367 (Aug. 1970), fig. 3.6-3.
13. PTR-738, p. 10.
14. Rasmussen Report (WASH-1400), app. 1, "Accident Definition," pp. 223, 245. This report does not mention the core drop, cascading control rod ejection, and autocatalytic accidents, however.
15. BNL-17608, A. Aronson et al., "Status Report on BNL Calculations of Anticipated Transients without SCRAM in Boiling Water Reactors" (Feb. 1973), fig. VIII-5.
16. Thompson and Beckerley, *Technology,* 1: 156, item no. 4.
17. A reactivity loss of 1.4% (reactivity units) was assumed.
18. WASH-1146, USAEC, *Water Reactor Safety Program Plan* (Feb. 1970; prepared by Water-Reactor Safety Program Office, Idaho Nuclear Corp., National Reactor Testing Station), p. III-102.
19. ANCR-1095, AEC R&D report, T. G. Odekirk, "Detailed Test Plan Report for PBF Test Series PCM-20: The Behavior of Unirradiated PWR Fuel Rods under Power-Cooling Mismatch Conditions" (Apr.

1974). GE report NEDO-10349, L. A. Michelotte, "Analysis of Anticipated Transients without SCRAM" (Mar. 1971). WCAP-8096, "Anticipated Transients without Reactor Trip [SCRAM] in Westinghouse Pressurized Water Reactors" (Apr. 1973). See also WASH-1146, p. III-102, and the PWR and BWR safety analysis reports for specific plants.

20. WCAP-8096, sec. 3.8.
21. "Westinghouse Reference Safety Analysis Report" (RESAR), rev. 3 (June 1972), pp. 5.4-17.
22. WCAP-8096, sec. 4.
23. Rasmussen Report (hereafter cited as Ras. Rpt.), 1: 208–33.
24. Ibid., 1: 186.
25. NEDO-10183, "Core Standby Cooling Systems for General Electric BWR Standard Plants (May 1970), p. 20. Also, it is of interest to note that the NRTS has stated that a blowdown in a BWR LOCA "could cause the water moderator level in the core region to rise. An increase in water level in the core region would result in a reactivity accident," i.e., a power excursion accident; AEC R&D report IN-1370, Annual Report SPERT Project (June 1970), p. 104.
26. Ras. Rpt., 1: 161–62.
27. WCAP-7422, "Westinghouse PWR Core Behavior Following a Loss of Coolant Accident" (Aug. 1971), sec. 4 and fig. 4-18.
28. For small breaks, the coolant pressure stays high (above 1,000 psi, which is about one half of the operating pressure) for 100 to 200 seconds, giving enough time, presumably, for core melting to occur. See RESAR, 7: 15.3-2; 15.4-101, 15.4-105. See also WCAP-7422, p. 4-11 and figs. 4-10 and 4-18, which indicate that core melting might occur within twenty seconds in a small-rupture LOCA, if the assumed SCRAM at two seconds after the start of the LOCA fails to occur. The SCRAM would greatly reduce the power level as discussed on pp. 4–11 and fig. 4-10 of WCAP-7422. The power-level reduction of the SCRAM explains why the fuel cladding temperature stops increasing after three seconds, as shown on fig. 4-18.
29. Ras. Rpt., 1: 131, 139, n. 10.
30. Ibid., 8: B-12.
31. Inferred from data in RESAR, 2: 4.4-40.
32. Ras. Rpt., 8: B-13.
33. Ibid., main vol., p. 246.
34. Inferred from data in RESAR, 2: 4.3-33 and 4.4-40; i.e., the "moderator temperature coefficient," and the potential for coolant temperature to drop in the core. It should be noted that WCAP-7422 does not plot results of small outlet breaks to give one an appreciation of the possibility for reactivity increases in this type of LOCA.
35. Ras. Rpt., 1: 124.
36. "Safety Evaluation by the office of Nuclear Reactor Regulation Supporting Amendment to License No. DPR-22 and Changes to the Technical Specifications Inoperable Control Rod Limitations," North-

ern States Power Company, Monticello Nuclear Generating Plant," USNRC docket no. 50-263 (Sept. 24, 1975).

37. Ras. Rpt., 1: 186.
38. Calculated from data in RESAR, 3: 5.4-17, table 5.4-1.
39. List of Documents Relating to the [1964–65] Reexamination of WASH-740, AEC Public Document Room, doc. no. 144-18, p. 12, memo from C. K. Beck to the Atomic Energy Commission on Draft Report to JCAE on the 1965 restudy of WASH-740.

Chapter Four

1. Thompson and Beckerley, *Technology,* 1: 5.
2. Ibid., chap. 11. R. L. Scott, "Fuel Melting Incident at the Fermi Reactor," *Nuclear Safety* 12, no. 2 (Mar./Apr. 1971): 123–34.
3. IDO-17028. Also AEC R&D report IDO-17281, R. K. McCardell et al., "Reactivity Accident Test Results and Analyses for the SPERT III E-Core: A Small Oxide-Fueled, Pressurized-Water Reactor" (Mar. 1969).
4. J. B. Yasinski and A. F. Henry, "Some Numerical Experiments Concerning Space-Time Reactor Kinetics Behavior," *Nuclear Science and Engineering* 22 (1965): 171–81. J. B. Yasinsky, "On the Use of Point Kinetics for the Analysis of Rod-Ejection Accidents," *Nuclear Science and Engineering* 39 (1970): 241–56.
5. WASH-1146, pp. III-94, 91–95.
6. A. F. Henry, "Status Report on Several Methods for Predicting The Space-Time Dependence of Neutrons in Large Power Reactors," *Space Dependent Reactor Dynamics,* Proceedings of a Specialist Meeting on "Reactivity Effects in Large Power Reactors," Oct. 28–31, 1970, Commission of the European Communities (Feb. 1972), p. 22; A. F. Henry, "Review of Computational Methods for Space-Dependent Kinetics," in *Dynamics of Nuclear Systems,* ed. D. L. Hetrick (University of Arizona Press, 1972), p. 10; and P. B. Parks et al., "Multidimensional Space-Time Kinetics Studies, Part II Experimental," to be published in *Nuclear Science and Engineering,* April 1976, see "Conclusions." See also W. M. Stacey, *Space-Time Nuclear Reactor Kinetics* (Academic Press, 1969), p. 4.
7. D. R. Ferguson and K. F. Hansen, "Solution of the Space-Dependent Reactor Kinetics Equations in Three Dimensions," *Nuclear Science and Engineering* 51 (1973): 189.
8. Ibid.
9. PTR-738.
10. Inferred from Montague I and II BWRs, Preliminary Safety Analysis Report, AEC docket no. 50-496-497, pp. 4.3-29, 15.1-143 through 15.1-155. See also NEDO-10527, C. J. Poane, "Rod Drop Analysis for Large Boiling Water Reactors," p. 3-1. Also USAEC Regulatory Guide 1.77 (May 1974).
11. IN-1370. See also NEDO-10527.

12. Henry, *Space Dependent Reactor Dynamics*, pp. 21, 24–25; NEDO-10527, p. 5-1; WCAP-7588, D. H. Risher, "An Evaluation of the Rod Ejection Accident in Westinghouse Pressurized Water Reactors Using Spatial Kinetics Methods" (Dec. 1971; rev. 1), p. 2–4.
13. IN-1370, "Large Core Dynamics," pp. 48–87. See also N. J. Diaz and M. J. Ohanian, "Reactor Dynamics Studies in Large Close-to-Critical Systems," in *Dynamics of Nuclear Systems*, ed. Hetrick, pp. 87–108.
14. IN-1370, p. 87; telephone conversation, N. J. Diaz (University of Florida) and R. E. Webb, May 2, 1975.
15. WAPD-TM-416, W. R. Cadwell et al., "WIGLE-A Program for the Solution of the Two-Group Space-Time Diffusion Equation in Slab Geometry" (1964).
16. Hetrick, *Dynamics of Nuclear Systems*, p. 174.
17. R. V. Meghreblian and D. K. Holmes, *Reactor Analysis* (McGraw-Hill, 1960), p. 351.
18. PTR-815, S. O. Johnson and R. W. Garner, "Large Core Dynamics Experimental Program Proposal" (Sept. 1966), internal report.
19. Ibid., pp. ii, 3, and 4. In the Phase II tests the core was to be large in one dimension, and in Phase III, large in two dimensions.
20. WASH-1146, pp. III-95, 96. In this regard, it is noted that Professor K. Hansen of MIT has asserted that the "neutron kinetics" of power excursion theory, i.e., neutron diffusion theory, has been verified in experiments performed at the Savannah River Laboratory in which the reactivity of a large core was raised to cause a power transient. He claims that these experiments adequately satisfy the objectives of the NRTS proposal for large-core dynamics experiments to verify neutron kinetics theory. (Nuclear safety conference-debate between K. Hansen, N. Rasmussen, et al. of MIT and R. E. Webb before the Massachusetts legislature's Committee on Nuclear Energy, Boston, Feb. 5, 1976). On the contrary, the Savannah River experiments, which did not involve power excursions but, rather, very slow rises in the power level, were qualified by those reporting the experiments as applying only to a nonpower excursion phenomenon called "delayed neutron holdback." They noted the second phenomenon of "super-prompt critical transients," i.e., power excursions, and stated: "Of necessity, these experiments have been restricted to delayed critical or to subcritical transients; thus, the delayed neutron holdback is the only neutronic space-time phenomenon accessible in the present experiments." (DB-MS-72-47, P. Parks et al., "Space-Time Kinetics Experiments Emphasizing Delayed Neutron Holdback"; and DP-MS-73-4, H. Dodd et al., "Space-Dependent Transient Reactor Experiments and Calculations at the Savannah River Laboratory," p. 2; see also Parks et al., "Multidimensional Space-Time Kinetics Studies.") Actually, qualitatively similar experiments were part of the NRTS proposal but were to be only preparatory to the proposed power excursion experiments. These preparatory tests were to consist of withdrawing a control rod on one side of a large core, enough to alter "greatly" the neutron density distribution in the

core as measured by a quantity called "flux tilt," which is the ratio of the neutron density on one side of the core relative to the opposite side. (A ratio of 1.0 would mean a balanced distribution.) In the context of the NRTS report (PTR-815) a "great" alteration would mean a tilt ratio changing from 1.0 at the start of a transient to 10,000 in the case of a control-rod-induced power excursion. (This is an example of a large-core neutronic effect.) In contrast, the Savannah River test experienced a flux tilt of only about 1.6, which is far from a test of flux tilt conditions in a DB-PEA; therefore, the Savannah River tests are by no means a definitive verification of neutron diffusion theory for DB-PEAs. (Strong changes in the neutron density may require neutron transport theory for accurate prediction.)

Moreover, no reactivity feedback effects were involved, as would be the case in a power excursion accident (DB-MS-72-47, p. 2). Also, the *measured* reactivity of the experiment, which is the primary quantity that sets the rate of power rise in any power transient, was used as input data for the calculation, to force it to accurately "predict" the transient, at least as to the rate of power rise. By so normalizing the calculations to fit the experiment, one cannot learn the degree to which the theory may have failed to predict the transient, including the time-varying flux tilt ratio. This, of course, further reduces the usefulness of the Savannah test. Finally, the reactor used in the Savannah experiments used deuturium oxide (heavy water), not ordinary water as is used in U.S. power plants, which is another difference. This difference affected the "largeness" of the reactor relative to the need to verify *large-core* neutronic effects. Though the size (diameter) of the Savannah River reactor was about the same as the cores of today's large PWRs and BWRs (light water reactors), the heavy water allowed the neutrons released in a fission to diffuse over greater distances before being absorbed, so that, equivalently, the Savannah test reactor represented only about one half the diameter of a large LWR core from the standpoint of investigating the validity of neutron diffusion theory in transient situations. Since the larger the core, the much more pronounced are the predicted large-core neutronic effects, such as flux tilt, and the greater, therefore, is the need to verify power excursion theory experimentally, the Savannah heavy water reactor would have had to have been double in size for it to have been applicable to LWRs. Overall, the experiments cited by Professor Hansen are not verification of DB-PEA theory for PWRs and BWRs.

21. See also the opinion of A. Henry in BNL-50117 (T-497), *Proceedings of the Brookhaven Conference on Industrial Needs and Academic Research in Reactor Kinetics* (Apr. 8/9, 1968; official use only), pp. 433, 453.
22. NRC letter from C. P. Jupiter to R. E. Webb, dated Dec. 24, 1975, forwarding the PTR-815 document.

23. This Phase II report is mentioned in WASH-1146, p. III-96.
24. Nuclear safety conference, Mass. legislature (recorded).
25. Telephone conversations, P. Parks and R. Webb, Feb. 8 and 9, 1976.
26. PTR-815, p. 8. Indeed, one million mesh points are "too coarse for many reactor problems," according to C. M. Kang and K. F. Hansen ("Finite Element Methods for Reactor Analysis," *Nuclear Science and Engineering* 51: 456). To reduce the number of mesh points, Kang and Hansen have investigated more sophisticated calculational methods than the "finite difference" method presently being relied on; but the more sophisticated "finite element" method has not been shown to be practical for DB-PEA problems, which will have spatial and time-varying core properties that may require a fine mesh in any event. (See A. F. Henry in *Dynamics of Nuclear Systems,* pp. 11–14.) This matter of required mesh-point number is not adequately treated in the open literature relative to accuracy requirements of practical problems—that is, DB-PEA predictions.
27. *Space Dependent Reactor Dynamics,* p. 26.
28. K. Hansen, MIT, nuclear safety conference, Mass. legislature. See also Ferguson and Hansen, "Solution," p. 202.
29. WCAP-7588, p. 5-1.
30. Diaz and Ohanian, "Reactor Dynamics Studies," pp. 105–6.
31. Ibid., p. 107.
32. To be published in *Nuclear Science and Engineering.*
33. It should be noted that neutron dynamics specialist A. Henry of MIT commented in 1969 on the neutron wave experiment that "I expect diffusion theory to fail because I don't think one should analyze this [experiment] except with a transport theory approach," since, said he, "the experiment creates sort of transport situation which is somewhat extreme" (BNL-50117, p. 262). Henry still feels the experimental results are "fairly uninterpretable," notwithstanding the preliminary two-dimensional diffusion theory analysis of the results by Parks (nuclear safety conference, Mass. legislature).
34. IN-1366, M. J. Ohanian and N. J. Diaz, "Final Report on Research on Neutron Pulse Propagation in Multiplying Media" (Idaho Nuclear Corp., Nov. 1969), p. 122.
35. Diaz and Ohanian, "Reactor Dynamics Studies," p. 95.
36. IN-1370, p. 87; and IN-1366, p. 171.
37. S. Kaplan et al., "Applications of Synthesis Techniques to Problem Involving Time Dependence," *Nuclear Science and Engineering* 18 (1964): 163–76; WAPD-MRP-113, *Shippingport PWR Technical Progress Reports,* pp. 84–87, 90, 108–9.
38. Thompson and Beckerley, *Technology,* 2: 66.
39. AEC R&D report IN-ITR-113, R. W. Miller, "The Effects of Burnup on Fuel Failure: I. Power Burst Tests on Low Burnup UO_2 Fuel Rods" (July 1970; a "limited distribution" document), pp. 49–51. See also IN-1370, pp. 26–27.
40. Unpublished report of "[T]he results of four power-burst tests on fuel rods with elevated burnups (13,000 to 32,000 MWD/MTU),"

Idaho Nuclear Corp., NRTS, pp. 24–27. See also IN-1313, J. E. Grund et al., "Subassembly Test Program" (Aug. 1969), pp. 62–90.
41. The report was sent to the AEC as a "preliminary copy" but was subsequently recalled by the test laboratory; letter from H. P. Pearson of Idaho Nuclear Corp. to S. A. Szawlewicz of the AEC, June 22, 1971. The title was to be "The Effects of Burnup on Fuel Failure: II. Power Burst Tests on Fuel Rods with 13,000 and 32,000 MWD/MTU Burnup," IN-ITR-118. A revised copy was to be sent to the AEC "soon," according to Pearson's letter; but evidently it was never sent.
42. R. L. Johnson et al., "Nuclear Safety Study: Fuel Behavior," *Nuclear News* (Aug. 1970), pp. 67–68. The failure is reported as 95 cal/gm, whereas the test report gives 85 cal/gm.
43. Ibid.
44. D. T. Aase, "Response of UO_2-Zircaloy Clad Fuel To Rapid Transients: Review of SPERT and TREAT Data" (Battelle Pacific Northwest Laboratories, Richland, Wash., Aug. 1973), p. 3.
45. Johnson, "Nuclear Safety Study," p. 68.
46. IN-ITR-116, L. J. Seifken, "The Response of Fuel Rod Clusters to Power Bursts" (May 1970; limited distribution report), pp. 34–39.
47. Unpublished report (IN-ITR-118).
48. USAEC WASH-1250, "The Safety of Nuclear Power Reactors (Light Water-Cooled)" (July 1973; final draft, July 16, 1973), pp. 7–20.
49. Letter from W. V. Johnston (NRC) to R. E. Webb, Sept. 26, 1975. The 95 cal/gm failure value quoted in the *Nuclear News* article refers to the same fuel rod failure specimen which the limited distribution test report said failed at 85 cal/gm.
50. 1. "Light-Water Reactor Fuel Behavior Program Description: RIA [Reactivity Initiated Accidents, i.e., PEAs] Experiment Requirements (Aerojet Nuclear Co., July 1, 1975), which contends: "Chapter 2 of this document surveys previous RIA fuel experiments. The relevant available data was generated in the SPERT and TREAT experiments" (p. 2). 2. Aase, "Response to Rapid Transients." 3. Johnson, "Nuclear Safety Study."
51. Telephone conversation, W. V. Johnston and R. E. Webb, Oct. 3, 1975.
52. L. C. Schmid, "A Review of Plutonium Utilization in Thermal Reactors, *Nuclear Technology* 18 (May 1973): 78.
53. M. Willrich and T. B. Taylor, *Nuclear Theft: Risks and Safeguards* (Ballinger, 1974).
54. IN-ITR-117, W. G. Lussie, "The Response of Heterogeneous Mixed-Oxide Fuel Rods to Power Bursts" (Idaho Nuclear Corp., Sept. 1970; limited distribution report).
55. IN-ITR-114, W. G. Lussie, "The Response of Mixed-Oxide Fuel Rods to Power Bursts" (Apr. 1970; limited distribution report), pp. 20–23.
56. "Power Burst Facility Test Program Plan," preliminary draft (was to be issued as IN-1434) (Idaho Nuclear Corp., NRTS June 1971), p. 3-18. E. Feinauer et al., "PBF Test Program Outline," preliminary

draft copy (was to be issued as IDO-17298), (Idaho Nuclear Corp., undated).
57. See Kouts, "AEC's Objectives, Plans, and Schedules," which show the extent of present facilities.
58. ANCR-1095, pp. 35–36.
59. Letter from W. V. Johnston, Sept. 26, 1975, in which it was stated that the original PBF test plans "assumed a cluster [multi-rod] testing capability which PBF cannot achieve." Compare the burn-up values of the "PBF Test Program Outline," IDO-17298, sec. 5.3.2 with those of "RIA Experiment Requirements," p. 64.
60. *U.S. Code of Federal Regulations,* chap. 10, pt. 50, app. A, no. 28.
61. WASH-1270, "Anticipated Transients without SCRAM for Water-Cooled Power Reactors" (Sept. 1973), USAEC, p. 66.
62. ANCR-1095, pp. 18, 21.
63. Ibid., p. 35.
64. RESAR, pp. 15.4-52 through 58; WCAP-8096, pp. 3.3-1 through 3; and WASH-1270, p. 63.
65. ANCR-1095, pp. 12–24.
66. Ibid., p. 18.
67. NEDO-10174, G. J. Scatena, "Consequences of a Postulated Flow Blockage Incident in a Boiling Water Reactor" (General Electric Co., May 1970).
68. Ibid., p. 28.
69. Ibid., pp. 32–38.
70. Ras. Rpt., 8: B-13.
71. Aerojet Nuclear Co., "Small Cluster Test Program Requirements Document" (Dec. 15, 1974), p. 62.
72. ANCR-1095, p. 109.
73. IN-1434; see n. 56 above.
74. "Power Burst Facility Test Program Plan" (final draft; was to be published as ANCR-1012), (Aerojet Nuclear Co., Mar. 1972).
75. WASH-1250, p. 7-R-2, item no. 26.
76. H. L. Coplen and L. J. Ybarrondo, "Loss-of-Fluid Test Integral Test Facility and Program," *Nuclear Safety* 15, no. 6 (Nov./Dec. 1974): 676–90.
77. APS Rpt., secs. 5B, 5C, 6D, 6F, and app. 1C; and Ford and Kendall, *Assessment of the ECCS Rulemaking.* See also, R. E. Webb, "Testimony on the AEC's Acceptance Criteria for Emergency Core Cooling Systems for Light-Water-Cooled Nuclear Power Reactors" (June 10, 1972), exhibits 1090 and 1091 in the AEC's ECCS hearing, July 12 and 13, 1972; see transcript, those dates (AEC docket no. RM-50-1).
78. Telephone conversation, G. Brockett, Inter-Mountain Technology Inc., Idaho Falls, Idaho, and R. E. Webb, Apr. 28, 1975.
79. APS Rpt., p. S62.
80. Thompson and Beckerley, *Technology,* 2: 804; compare Indiana Pt. 2 with Browns Ferry.
81. R. E. Webb, Testimony in the Matter of a LOCA in the Proposed Shoreham Power Station, comments 12 and 18 (Dec. 19, 1971), AEC

docket no. 50-322. (See also R. Webb, ECCS testimony.) Also, NEDO-10329, B. C. Slifer and J. E. Hench, "Loss-of-Coolant Accident and Emergency Core Cooling Models for General Electric Boiling Water Reactors" (Apr. 1971), p. A-12.
82. APS Rpt., pp. S62–68 and S77–78.
83. Kouts, "Nuclear Safety Research Objectives, pp. 128–29.
84. PTR-815, p. 2.
85. H. Bethe, "The Necessity of Fission Power," *Scientific American* (Jan. 1976), pp. 21, 25.
86. WASH-1146, p. III-89.
87. IDO-17028, p. 46.
88. Thompson and Beckerley, *Technology,* vol. 1, chap. 11, secs. 3.4 and 3.12. See also sec 3.11 on the SL-1 accident.
89. J. R. Dietrich, "Experimental Determination of the Self-Regulation and Safety of Operating Water-Moderated Reactors," *Proceedings of the International Conference on the Peaceful Uses of Atomic Energy* (1955), 13: 89. See also A. W. Kramer, *Boiling Water Reactors* (Addison-Wesley, 1958), p. 74.
90. Kramer, *Boiling Water Reactors,* p. 75.
91. BMI-1910, p. C-16.
92. Thompson and Beckerley, *Technology,* 1: 623.
93. Kramer, p. 64.
94. IDO-17028, p. 46.
95. A. H. Spano, "Self-Limiting Power Excursion Tests of a Water-Moderated Low-Enrichment UO_2 Core," *Nuclear Science and Engineering* 15 (1963): 37–51; IDO-17028, p. iv.
96. IDO-17281; R. K. McCardell et al., "Reactivity Accident Test Results and Analysis for the SPERT III E-Core" (March 1969).
97. Ibid., p. 85.
98. PTR-815, pp. 17–19 and 30.
99. Kouts, "AEC's Objectives, Plans, and Schedules," p. 130.
100. IDO-16879, A. A. Wasserman et al., "Power-Burst Facility (PBF) Conceptual Design" (June 1963), pp. 29–30.
101. WASH-1250, pp. 7–20.

Chapter Five

1. AEC Atomic Safety and Licensing Board hearing on the Shoreham BWR, docket no. 50-322 (Nov. 3, 1971), transcript p. 11,408 (cross-examination of P. W. Ianni by R. E. Webb).
2. AEC letter from S. H. Hanauer to R. E. Webb, Dec. 9, 1974. See also WASH-1270, p. 49.
3. AEC hearing on ECCS, RM-50–1 (April 11, 1972), transcript p. 8016 (cross-examination of S. Hanauer by R. E. Webb).
4. WASH-1270, p. 86.
5. At any rate, WASH-1270 does not require a different design for the control rods and their drive mechanisms (see p. 86).
6. *Nuclear Safety* 15, no. 2 (Mar./Apr. 1974): 210–211.

7. Vermont Yankee Nuclear Power Corp. letter to AEC, Nov. 14, 1973 (VYV-3071), AEC docket no. 50-271, p. 3.
8. AEC letter to Vermont Yankee Nuclear Power Corp., Dec. 27, 1973, app. A.
9. Incidents: Shippingport, personal observation, Nov. 1964; Fermi, see Scott, "Fuel Melting Incident," p. 123; Millstone Point, see *Nuclear Safety* 12, no. 6 (1971): 619 (see also, steam bypass failure-to-open incident at Millstone Point, Sept. 23, 1971, AEC public document room); SCRAM inoperative, WASH-1270, p. 49; and Dresden, USAEC ROE 71-4 (Mar. 31, 1971), "Reactor Safety Operating Experiences."
10. WASH-1270, p. 65.
11. RESAR, p. 5.5-62.
12. APS Rpt., pp. S39 and S91.
13. Kendall, "Nuclear Power Risks," p. 15. Also, R. E. Webb, "Operating Experiences of Steam Generators and Heat Exchangers in Nuclear Power Plants and the Implications for the Midland Plant, Units 1 and 2," testimony before AEC Atomic Safety and Licensing Board, docket nos. 50-329 and 50-330 (Sept. 21, 1971).
14. RESAR, pp. 5.5-17 and 5.2-25.
15. A manager of Westinghouse has stated: "In regard to your question concerning the rupture of steam generator tubes coincident with a LOCA, Westinghouse firmly believes that such an occurrence is extremely improbable.... Analyses done by Westinghouse show that the tubes, even if extensive corrosion had occurred, would maintain their integrity in a LOCA. Westinghouse, in addition, recommends steam generator operation conditions, water chemistry, controls, and inspection practices which assure that corrosion of steam generator tubes will be minimized during operation." Letter from R. A. Wieseman to R. E. Webb, Nov. 4, 1975. However, no documentation was cited.
16. Bailly BWR PSAR, p. 14.6-3.
17. NEDO-10527, figs. 3-2 and 3-4.
18. Montague PSAR, pp. 4.3-28, 4.3-29, 15.1-153, and 15.1-154.
19. Fermi; see Scott, "Fuel Melting Incident," p. 129.
20. See app. 2, no. 2.
21. Ras. Rpt., 3: 30–31.
22. Ibid., p. 121.
23. ORNL-NSIC-64, "Abnormal Reactor Operating Experiences: 1966–1968" (Nuclear Safety Information Center, Oak Ridge National Laboratory, 1969), p. iv; ORNL-NSIC-69, "Safety-Related Occurrences in Nuclear Facilities as Reported in 1967 and 1968," pp. 2–3; ORNL-NSIC-103, "Abnormal Reactor Operating Experiences: 1969–1971," p. iv.
24. WASH-1208, "Status of Central Station Nuclear Power Reactors, Significant Milestones" (Jan. 1975), USAEC.
25. See app. 2, no. 11.
26. This author has urged this undertaking in testimony before the

AEC's licensing hearing on the Shoreham BWR in 1971 and before the AEC's ECCS safety hearing in 1972. However, the Shoreham testimony was completely ignored by the AEC's Atomic Safety and Licensing Board, when the board granted the Shoreham construction permit (AS&LB's Initial Decision, Shoreham Nuclear Power Station, docket no. 50-322 [Apr. 12, 1973], pp. 24–27; Testimony, n. 81 above). Also, urging of the undertaking and related technical points were stricken from the author's ECCS testimony by the AS&LB (ECCS transcript, July 13, 1972).

Chapter Six

1. Ras. Rpt.
2. Ibid., app. 6, pp. 77, 80, and 86.
3. Nuclear debate between Prof. N. Rasmussen (MIT) and Prof. David R. Inglis (University of Massachusetts) at Marlboro College, Marlboro, Vt., Apr. 14, 1975. Questioning from the floor by R. Webb (recorded).
4. Ras. Rpt., summary, p. 25.
5. Ibid., 1: 223, 245.
6. Ibid., 1: 198, 200, 208, and 234.
7. Ibid., 1: 197, 245; 9: 47.
8. Personal observation (Nov. 1964). See app. 2, no. 3.
9. DC-10 crash, Mar. 4, 1974, Paris, France: 345 persons killed.
10. Ras. Rpt., 1: 28, 186; 9: 47–49.
11. Ibid., 1: 186.
12. Ibid., p. 198.
13. Ibid., 6: 9, PWR-2 release category; and 5: 115.
14. WASH-740, pp. 35–36, 41–42.
15. Thompson and Beckerley, *Technology,* vol. 2, Chap. 18, esp. sec. 3.3.
16. W. E. Browning et al., "Release of Fission Products during In-Pile Melting of UO_2," *Nuclear Science and Engineering* 18 (1964): 151–62.
17. Ras. Rpt., 7: D-11 through D-14. See also "Critique by the General Electric Co. of the Discussion Prepared by the Union of Concerned Scientists . . . Concerning the Possible Consequences of a Reactor Accident," AEC ECCS hearing, docket no. RM-50-1 (Aug. 9, 1973), p. 12.
18. WASH-1250, pp. 7–17. Browning "Release of Fission Products," p. 152. IDO-17049, T. R. Wilson et al., "An Engineering Test Program to Investigate a Loss of Coolant Accident" (Oct. 1964).
19. Ras. Rpt., main vol., p. 245.
20. WASH-740, p. 23.
21. Ras. Rpt., app. 7, "Release of Radioactivity in Reactor Accidents," pp. 8–13. The main author of app. 7, R. L. Ritzman, has confirmed to this author that the 6% Sr 90 release assumption is not applicable to severe nuclear excursion accidents, i.e., PEAs.

22. WASH-740, pp. 10, 35.
23. Inferred from assumptions of deposition velocities made in the Ras. Rpt., app. 6, p. 20, compared with correlation with particle size in WASH-740, p. 50, and WASH-740's assumption of one-micron size, p. 8.
24. Compare SR 90 contamination limits in RAS. Rpt., app. 6: p. 66, with WASH-740, p. 41.
25. Ras. Rpt., app. 6, pp. 77, 80, 86; WASH-740, p. 100.
26. Compare Ras. Rpt., app. 6, p. 5, with WASH-740, p. 25.
27. Ras. Rpt., app. 6, pp. 64, 101–2.
28. HW-69561 AEC R&D report, "The Consequences of Accidental Releases during Shipments of Radioactive Cesium and Strontium," rev. and del.) (June 1961), p. 18. D. S. Smith, "Interim Protective Action Levels," AEC-sponsored Regional Workshop Seminar on State Emergency Planning in Relation to Licensed Nuclear Facilities, Oak Ridge, Tenn. (Sept. 1972); and P. G. Voilleque, "Dose Action Levels for Accidental Radiation Exposure of the General Public," Radiation Protection Standards: Quo Vadis, Sixth Annual Health Physics Society Topical Symposium, Richland, Wash., Nov. 1971 (Health Physics Society, 1972), 1: 183–204, which are refs. no. 38 and 39 of Ras. Rpt., app. 6.
29. H. M. Parker and J. W. Healy, "Environmental Effects of a Major Reactor Disaster," *Proceedings of the International Conference on the Peaceful Uses of Atomic Energy,* Geneva, 1955 (United Nations, 1956), 13: 106–9.
30. Ras. Rpt., app. 6, p. 102, to wit: "J. W. Healy, Los Alamos Scientific Laboratory, internal publication."
31. Letter from J. W. Healy, Los Alamos Scientific Lab., to R. E. Webb, May 23, 1975, forwarding Healy's "Nuclear Reactor Accident: Damage Assessment" (draft: Jan. 1966), section on Agriculture—Long Term Restrictions.
32. Letters from J. W. Healy to R. E. Webb, May 23, 1975, and May 5, 1975.
33. Ras. Rpt., app. 7, p. 111; app. 6, p. 9.
34. Ibid., app. 6, p. 68.
35. Ibid., app. 1, p. 197.
36. Ibid., app. 5, 14–20, esp. pp. 18–19; and app. 1, p. 209. See also app. 1, p. 143, in which the spontaneous reactor vessel rupture accident is assumed to be "comparable [to a] large LOCA." See also app. 5: p. 12.
37. Ibid., app. 5, pp. 109–72 (attachment 1).
38. The Ras. Rpt. claims that "Considerable effort was spent in trying to identify possible accidents in which a release larger than that of category 1 might be produced." Yet, of the "processes" considered, none included prompt containment failure; main vol. p. 119. See also n. 36, above.
39. Ibid., app. 1, pp. 223, 245.
40. Ibid., app. 5, pp. 84–85, 100–4.

41. Ibid., ap. 1, p. 142.
42. Ibid., p. 196.
43. Ibid., fig. I-21.
44. Ibid., pp. 208–33.
45. Ibid., app. 5, pp, 29, 35 (tables V-9, V-14).
46. Ibid. Compare table V-6 with V-9.
47. NRC meeting between H. C. Kouts, Director of Regulatory Research, S. Levine, et al., and R. E. Webb, Sept. 30, 1975, Wash. D.C.
48. Ras. Rpt., app. 5, p. 103; app. 1, p. 255–56.
49. Montague PSAR, table 5.2.1.
50. NEDO-10349, pp. 12–13.
51. Ibid., p. 42.
52. Ras. Rpt., app. 5, p. 103.
53. Ibid., p. 35 (table V-14).
54. Ibid., pp. 38, 100, 103, 104.
55. Ibid., p. 38.
56. Ibid., app. 5, pp. 102–4; app. 8, pp. 49–50.
57. Ibid., app. 5, p. 68.
58. Ibid., app. 1, p. 227.
59. Ibid., pp. 75–77. The probability of a steam explosion occurring which ruptures the containment is estimated at 1%; see app. 5, pp. 124–25, 155.
60. Ibid., app. 8; p. 2.
61. Ibid., pp. B-9 throught B-13.
62. Ibid., p. B-9.
63. Ibid., app. C.
64. ANL-7742, "Reactor Development Program Progress Report" (Sept. 1970), pp. 151–54. This report contains photographs of small-scale test explosions.
65. Ras. Rpt., p. 246.
66. Ibid., p. 230.
67. S. Levine, NRC-Webb meeting, Sept. 30, 1975.
68. ORNL-NSIC-69, p. 2.
69. PTR-815, p. 1.
70. Ras. Rpt., app 3, p. 121.
71. A Westinghouse employee, Winston Little of Hanford Engineering Development Laboratory, has acknowledged that he was the sole author of this appendix of the Ras. Rpt. Telephone conversations between W. Little and R. E. Webb, Aug./Sept. 1974.

Chapter Seven

1. List of Documents Relating to the Reexamination of WASH-740.
2. Ibid., nos. 111-20, 111-11, 111-17.
3. Ibid., no. 147–1.
4. Ibid., nos. 1a-7, 78-3, 113-13, 111-22, 106-16.
5. Ibid., nos. 92-5, 113-13, 71-1.

6. Ibid., nos. 180-31 through 180-33 (compare *C* value with implicit value of HW-69561, p. 18). See also no. 41-4.
7. Ibid., nos. 144-31 through 144-32.
8. Ibid., no. 155-3.

Chapter Eight

1. APS Rpt.; *Physics Today* 28, no. 7 (July 1975); 38–43, 59, 80.
2. APS Rpt., p. S5.
3. Ibid.
4. IN-ITR-113, p. 51.
5. APS Rpt., p. S28.
6. Ibid., p. S59.
7. Ibid., p. S53 (emphasis added).
8. Kouts, "AEC's Objectives, Plans, and Schedules," p. 48.
9. PBF Conceptual Design, IDO-16879, pp. 2–3.
10. PTR-738, pp. 107–8.
11. IDO-16879, p. 3.
12. APS Rpt., pp. S34, 43, 44.
13. Ibid., p. S4.
14. Ibid., pp. S62–65.
15. *Nuclear Safety* 15, no. 2 (Mar./Apr. 1974): p. 212.
16. APS Rpt., pp. S50–51.
17. Ibid., table 10, p. S47; table 11, p. S48; table 12, p. S49.
18. Ibid., p. S107.
19. Ibid, p. S109.
20. Ibid., Options, pp. S77–S78.

Chapter Nine

1. WASH-1208; Thompson and Beckerley, *Technology,* vol. 1, app. 1.
2. PTR-738, p. 108.
3. Ibid.
4. Ibid.
5. This author learned of the existence of the PTR-738 report when he sought to find out what PEA studies have been done by the NRTS, as it was this laboratory that conducted the early power excursion experiments on small reactors.
6. PTR-738, p. 3.
7. WASH-1208.
8. PTR-738, p. 2.
9. Unpublished report, IN-ITR-118, p. 1; and telephone conversation with N. Diaz, May 2, 1975. Also, phases II and III of the large-core kinetics program, mentioned in WASH-1146, pp. III-93–96, were never carried out.
10. *Atomic Power Safety,* USAEC series of booklets, *Understanding the Atom* (June 1964), pp. 35, 37.
11. Letter from R. E. Webb to NRC chairman, Aug. 17, 1975.
12. Letter from J. Kout to R. E. Webb, Sept. 23, 1975.

13. USNRC NUREG-75/058, *Reactor Safety Research Program* (June 1975), pp. 40–41.
14. J. R. Dietrich, "The Reàctor Core," Thompson and Beckerley, *Technology,* vol. 1, sec. 10.3.3.
15. Rasmussen-Inglis debate; Fox, Kadak: Nuclear debate, Westledge School, West Simsbury, Conn., March 7, 1975; Coughlin, Bailey: Open Forum debate, Indiana Univ. S.E., Sept. 21, 1974—all recorded.
16. I. A. Forbes et al., "The Nuclear Debate: A Call To Reason," Energy Research Group, 45 Hardings St., Belmont, Mass. 02178.
17. Telephone conversation between H. Kouts (NRC) and R. E. Webb, July 1975.
18. Telephone conversation with W. V. Johnston, Oct. 3, 1975.
19. Incidentally, the PTR-738 power excursion calculations may have excluded a mitigating factor, called prompt moderator heating, (PMH), as the report is not clear on this. This would mean that for some of the excursions calculated, the severity might have been overpredicted; but not drastically. However, the PMH effect would be insignificant for the worst case calculated (an excursion occurring when the coolant is cold); and furthermore, the initial conditions assumed for those excursions that were calculated were far from the worst case possible relative to the initial power level and reactivity. Also, the reactivity feedback due to PMH may be insignificant in PWRs, due to the boron, or positive (worse).
20. *Nuclear Safety* 15, no. 3 (1974): 256.
21. Thompson and Beckerley, *Technology,* 1: 176.
22. Ibid., pp. 684–85.
23. Ibid., p. 623.
24. Ibid., pp 625–33. See also app. 2, no 1, herein.
25. Testimony of the AEC Regulatory Staff on Interim Acceptance Criteria for Emergency Core Cooling Systems for Light-Water Power Reactors (Jan. 27, 1972), pp. 3-51 through 3-56. See also APS Rpt., pp. S68–69.
26. However, the industry has since argued persuasively that the semiscale test was not a fair simulation and that other ECCS systems would still provide the necessary cooling, so that the theoretical flaw uncovered by the test can be corrected without seriously affecting ECCS performance predictions. (Westinghouse testimony before the AEC ECCS hearing, docket no. RM-50-1 [Mar. 23, 1972], pp. 39–45; and AEC testimony, no. 25, above.) Subsequent testing with improved simulation showed that the coolant stays in the vessel, as is now predicted with the improved theory. (J. O. Zane, U.S. Congress, *Hearings before the Joint Committee on Atomic Energy,* 93d Cong., 2d sess. [Jan. 1974], pt. 2, vol. I, p. 460.)
27. Isaac Newton, *The Mathematical Principles of Natural Philosophy,* trans. by Andrew Motte (London, 1729); Mr. Cote's preface (1968, Dawsons of Pall Mall, London).
28. Ibid., 2: 392–93.
29. PFES, 6: 38–278.
30. NUREG-75/058.

31. PTR-738; WASH-1146.
32. Aerojet Nuclear Co., "Small Cluster Test Program," p. 6.

Chapter Ten

1. "Breeder Reactors," in AEC booklet, *Understanding the Atom* (1971).
2. WASH-1184, "Cost-Benefit Analysis of the U.S. Breeder Reactor Program" (Jan. 1972), USAEC, pp. 34–39. See *Hearings*, app. 9.
3. ANL-7657, C. Kelber et al., "Safety Problems of Liquid-Metal-Cooled Fast Breeder Reactors" (Feb. 1970); R. E. Webb, "Some Autocatalytic Effects during Explosive Power Transients in Liquid Metal Cooled, Fast Breeder, Nuclear Power Reactors (LMFBRs)" (Ph.D. diss., Ohio State University, 1971); obtainable from University Microfilm, Inc., Ann Arbor, Mich., no. 72-21, 029; see *Dissertation Abstracts*, 33, no. 2 (Aug. 1972): 754B–755B. See also chap. 1, nn. 1, 2.
4. See "Safety Criteria," no. 7, of K. P. Cohen and G. L. O'Neill, "Safety and Economic Characteristics of a 1,000 MWe Fast Sodium-Cooled Reactor Design," in ANL-7120, *Proceedings of the Conference on Safety, Fuels, and Core Design in Large Power Reactors* (Oct. 11–14, 1965), pp. 185–86.
5. Substituting Pu for "fission products" in fig. 14 of WASH-740, p. 70, and table 3, p. 100. That this estimate is plausible is supported by the fact that the nuclear industry must show that the public is not harmed if only 1/10 of a gram of Pu is released. See PFES, 2: 4.2-161.
6. Willrich and Taylor, *Nuclear Theft*, p. 26.
7. ANL-7520, *Proceedings of the International Conference on Sodium Technology and Large Fast Reactor Design* (Nov. 7–9, 1968), pt. 2, p. 351. See also ORNL-NSIC-74, B. P. Fish et al., "Calculation of Doses Due to Accidentally Released Plutonium from an LMFBR" (Nov. 1972), p. 5.
8. Willrich and Taylor, *Nuclear Theft*, p. 26.
9. Ras. Rpt., app. 6, p. 66.
10. A R. Tamplin and T. B. Cochran, "Radiation Standards for Hot Particles," Natural Resource Defense Council, 1710 N St., N.W., Washington, D.C. 20036.
11. Webb, "Some Autocatalytic Effects," p. 1.
12. ANL-7532, G. H. Golden, "Elementary Neutronics Considerations in LMFBR Design" (Mar. 1969), pp. 55–56.
13. Webb, "Some Autocatalytic Effects," p. 1.
14. PFES, 2: 42–113.
15. GEAP-10010-31, AEC R&D report, SEFOR Development Program, 31st and Final Report (Feb. 1972), pp. 5-1–5-5, 6-68, 2-1, 2-2. See also, GEAP-13929, SEFOR Experimental Results and Applications to LMFBRs" (Jan. 1973), pp. 1-1, 2-11, 4-1. See also *Hearings*, p. 181.
16. HEDL-TME-71-34, D. Simpson et al., "Assessment of Magnitude

and Uncertainties of Hypothetical Accidents for the FFTF" (Mar. 27, 1971), p. 38.

17. GEAP-10010-31, p. 5-1.
18. ANL-7657, p. 98.
19. WASH-1509, *LMFBR Demonstration Plant Environmental Statement* (Apr. 1972), USAEC, p. 54; AEC comments on the statement by Dr. Barry J. Smernoff, entitled "Underground Siting of the LMFBR Demonstration Plant: A Serious Alternative." See p. 132 of *Hearings*. See also PFES, 2: 4.2-154.
20. PFES, 2: 4.2-144.
21. Webb, statement on LMFBR Demonstration Plant, *Hearings*, p. 185.
22. PFES, 2: 4.2-160. The AEC's estimate of the explosion is 2,200 megawatt-seconds of energy, which can be converted to 1,100 lb. TNT-equivalent (1 lb. of TNT = 2 megawatt-secs.).
23. ANL-7657, p. 63.
24. Webb, "LMFBR Explosion Hazard," pp. 134–35. This estimate was made on the basis of assuming that a 5-lb. TNT-equivalent sodium vapor explosion drives the equivalent of 5% of the core fuel back into the core, which would raise the reactivity at the rate of about 2,200 $/sec. (1$, the threshold for prompt critical, equals .3% reactivity in plutonium-fueled LMFBRs.) This rate is derived by multiplying the speed that such a mass of fuel could attain by the explosion by the rate of reactivity rise per meter of length, which is 5% reactivity, or 17$, divided by the length of the core, assumed to be one meter. The mass of the core is 23,000 kilograms. The 5% reactivity would roughly be the reactivity gain if the fuel mass were fully inserted into the core. Meyer and Wolfe calculated 650 lb. of TNT-equivalent explosion for a 100 $/sec reactivity rate (ANL-7120, p. 681). By ratioing from this rate, 2,200 $/sec would yield a 14,000 lb. TNT explosion. However, the Bethe-Tait theory suggests a "3/2 power" rule, which would raise the estimate to 67,000 lb. TNT (Thompson and Beckerley, *Technology*, 1: 588–93).
25. Webb, "LMFBR Explosion Hazard," and "Some Autocatalytic Effects."
26. ANL/RAS 70-06, "LMFBR Accident Delineation and Recommended Program of In-Pile Safety Experiments" rev. (draft, Aug. 1970; rev. May 1971), Argonne National Laboratory, Reactor Analysis and Safety Division, *Project Report: Studies of LMFBR Safety Test Facilities*, vol. 2.
27. PFES, 2: 4.2-222.
28. Clinch River Breeder Reactor Plant, Preliminary Safety Analysis Report (Oct. 1975), app. E, pp. E.1-1, amendment 5.
29. W. J. McCarthy and D. Okrent, "Fast Reactor Kinetics," in Thompson and Beckerley, *Technology*, 1: 586–87; W. J. McCarthy et al., "Studies of Nuclear Accidents in Fast Power Reactors," *Proceedings of the International Conference on the Peaceful Uses of Atomic Energy* (1964), 12: 224.
30. ANL-7520, P. M. Murphy et al., "International Conference on So-

dium Technology and Large Fast Reactor Design" (Nov. 1968), pt. 2, pp. 356–57.

31. PFES, 6: 38-276–79. See this author's reply, ibid., pp. 280–90.
32. Ibid., 2: 4.2-155.
33. J. F. Jackson and J. D. Boudreau, "Postburst Analysis," preliminary report, (Argonne National Lab., Dec. 1972). Cited in PFES, 6: 38-281, 251; 2: 4.2-148.
34. PFES, 6: 38-281. The 1,160-lb.-TNT figure was inferred from the fuel temperature data in the Jackson-Boudreau report, pp. 61–64, allowing for spatial power peaking, with the aid of data by Hicks and Menzie, ANL-7120, p. 659, and adjusting for the mass.
35. USAEC Directorate of Licensing, "Safety Evaluation of the Fast Flux Test Facility" (Oct. 31, 1972), project no. 448, p. 1. The FFTF is not an excursion test reactor. Its purpose is to test fuels under normal conditions, pp. 106–7.
36. Ibid., p. 30; PFES, 2: 4.2-170.
37. PFES, 2: 4.2-147–49.
38. Ibid., pp. 4.2-144–51.
39. J. E. Boudreau and J. F. Jackson, "Recriticality Considerations in LMFBR Accidents," *Proceedings of the Fast Reactor Safety Meeting*, Beverly Hills, Calif., Apr. 2–4, 1974 (USAEC CONF-740401), p. 1265.
40. Ibid.; quoted in PFES, 2: 4.2-149.
41. Webb, "Some Autocatalytic Effects," chaps. 5 and 6.
42. Boudreau and Jackson, "Recriticality Considerations," p. 1266.
43. Ibid., p. 1276.
44. Ibid, p. 1279.
45. USAEC Division of Reactor Research and Development, "Discussion Regarding Postburst Analysis (Preliminary Report) by J. Boudreau and J. F. Jackson" (undated); transmitted by letter to R. H. Sandler by B. H. Schur (AEC), July 23, 1973, and to R. E. Webb by T. A. Nemzek (AEC), Oct. 19, 1973.
46. ANL/RAS 72-30, H. K. Fauske, "On the Mechanism of UO_2-Na [uranium oxide-sodium] Vapor Explosions" (Aug. 1972). See also *Nuclear Science and Engineering* 51, no. 2 (June 1973).
47. See n. 22 above, and compare with value in n. 4 (see p. 118 above).
48. PFES, 2: 4.2-153.
49. Ibid., pp. 4.2-152, 162, 227.
50. R. P. Anderson and D. R. Armstrong, "Comparison between Vapor Explosions Models and Recent Experimental Results," American Institute of Chemical Engineers, 14th National Heat Transfer Conference, Aug. 5–8, 1973, pp. 29–31. ANL/RDP-2, "Reactor Development Program Progress Report" (Feb. 1972), pp. 8.31–32. R. P. Anderson and D. R. Armstrong, "Laboratory Tests of Molten-Fuel-Coolant Interactions," *Transactions*, American Nuclear Society, 15, no. 1 (June 1972): 313. See also *Nuclear Safety* 16, no. 4 (1975): 438.

51. H. K. Fauske, "Some Aspects of Liquid-Liquid Heat Transfer and Explosive Boiling," *Proceedings of the Fast Reactor Safety Meeting*, pp. 992–1005.
52. Ras. Rpt., app. 8, pp. B-5–B-9.
53. PFES, 2: 4.2-153.
54. Fauske, ANL/RAS 72-30, figs. 3 and 9, and p. 7 (see n. 46 above).
55. Fauske, "Aspects of Heat Transfer," p. 998.
56. Ibid.
57. Ibid.
58. M. Amblard et al., "Out of Pile Studies in France on Sodium Fuel Interaction," *Proceedings of the Fast Reactor Safety Meeting*, pp. 911–12. See also n. 50 above.
59. Fauske, "Aspects of Heat Transfer," pp. 997–98. Anderson and Armstrong, "Comparison," pp. 39–40.
60. PFES, 2: 4.2-153, which cites a paper by J. F. Marchaterre et al. of ANL on "Experimental Studies of Liquid-Metal Fast Reactor Accident Conditions," International Conference on Engineering of Fast Reactors for Safe and Reliable Operation (Karlsruhe, Oct. 9–13, 1972), 2: 763–86, 773–75.
61. ANL has said that the TREAT tests are not "directly applicable"; ibid., p. 774.
62. For example, T. R. Johnson et al. of ANL reported experiments involving a few pounds of molten UO_2 dropped in sodium. No sodium vapor explosions were observed. These experiments were to resemble in a vague way the molten fuel-sodium interaction process that was assumed by Boudreau and Jackson for their sodium-vapor-explosion mode of fuel recompaction. However, the authors did not analyze the phenomena and conditions of the tests relative to expected LMFBR conditions and sodium explosion theories, in order to determine the possible conclusions one might draw about the meaning of these experiments with respect to the LMFBR explosion potential. On the other hand, the authors drew no such conclusion. Indeed, experiments similar to these produced "violent interactions," presumably explosions (see Amblard, "Out of Pile Studies").
63. Ras. Rpt., app. 8, pp. B-5–B-9. W. Zyszkowski, "On the Initiation Mechanisms of the Explosive Interaction of Molten Reactor Fuel with Coolant," *Proceedings of the Fast Reactor Safety Meeting*, p. 897.
64. G. Angerer et al., "Critical Discussion of Some Important Topics in Fast Reactor Safety Analysis," *Proceedings of the Fast Reactor Safety Meeting*, p. 1230.
65. See nn. 46 and 51 above.
66. ANL-7120, F. P. Hicks and D. C. Menzies, "Fast Reactor Maximum Accident," pp. 663–65.
67. Estimated from reactivity change rates in fig. 23 of J. E. Boudreau and R. C. Erdmann, "On Autocatalysis during Fast Reactor Dis-

assembly," *Nuclear Science and Engineering* 51 (1973): 206–22. The estimate assumes that if the reactivity rises above zero, it will do so at about the same rate at which it was predicted to drop.

68. Webb, "Some Autocatalytic Effects," chaps. 4 and 7, pp. 206–8. The author's crude estimates of the reactivity potential of the reduction of neutron streaming have since been supported by P. Kohler and J. Ligow, "Axial Neutron Streaming in Gas-Cooled Fast Reactors, *Nuclear Science and Engineering* 54 (1974): 357–60. These authors predict a reactivity effect in the range of 1% to 1.5% reactivity, which compares well with this author's range of .9% to 1.7% (chap. 4, p. 127). Remember that percent units of reactivity have special meaning. For example, the 9,000-lb.-TNT explosion estimate is based on a .6% autocatalytic reactivity effect due to streaming reduction. The gas-cooled reactor would be very similar to an LMFBR core without sodium; so the results of Kohler and Ligow are applicable.
69. Boudreau and Jackson, "Recriticality Considerations," p. 1271. The 10,000-lb.-TNT excursions are inferred from their calculated reactivity insertion rates for coherent fuel reentry. See also Webb, "Some Autocatalytic Effects," pp. 208–9.
70. Implosion reactivity is investigated in Webb, "Some Autocatalytic Effects." Significant autocatalytic implosive reactivity feedback has been calculated for design basis accidents in the LMFBR Demonstration Plant, which needs to be thoroughly studied. (See Clinch River Breeder Reactor PSAR, pp. F6.2-112, amendment 5.
71. PFES, 2: 4.2-147 through 4.2-148. See also Boudreau and Jackson, "Recriticality Considerations"; and J. F. Jackson et al., "Trends in LMFBR Hypothetical Accident Analysis," *Proceedings of the Fast Reactor Safety Meeting*, sec. 7. Also, the NRC's *Reactor Safety Research Program*, NUREG-75/058, includes theoretical and experimental investigations of the neutron streaming questions (pp. 26–27).
72. ANL-7120, pp. 269–70.
73. R. Avery, *Proceedings of the Fast Reactor Safety Meeting*, p. 379. See also J. B. van Erp and H. K. Fauske, ibid., p. 1432: "And once the analysis is followed to the point of partial or total core meltdown, many different scenarios are possible, including second and subsequent criticality, none of which can, in the present state of knowledge, be easily disproven with a sufficiently high confidence level to be able to exclude them."
74. PFES, 2: 4.2-146.
75. Avery, *Proceedings of the Fast Reactor Safety Meeting*, p. 380; van Erp and Fauske, ibid., p. 1433, who use the phrase "reasonably provable high limits"; R. B. Nicholson, ibid., p. 1665. See also Jackson et al., "Trends," sec. 10, pp. 1260–61.
76. Boudreau and Jackson, "Recriticality Considerations," pp. 1265–82, 1276.
77. *Hearings*, p. 189.

78. See fig. 25 of Boudreau and Erdmann, "Autocatalysis."
79. PFES, 2: 4.2-147–48.
80. *Report of the Cornell Workshops on the Major Issues of a National Energy Research and Development Program*, Cornell University, Sept./Oct. 1973 (rev. ed., Dec. 1973), pp. 183–84.
81. NUREG-75/058, secs. 4 and 6. The Safety Test Facility mentioned in this NRC document for "transient in-reactor tests" is not an integral core destruct test (see H J. C. Kouts, "The Future of Reactor Safety Research," *Bulletin of the Atomic Scientist*, Sept. 1975, p. 36).
82. ANL-7657, pp. 84–85.
83. ANL/RAS 70-06, p. III-14.
84. Letter from L. Baker (ANL) to R. Webb, Jan. 23, 1973.
85. Kouts, "Future of Reactor Safety Research," p. 36.
86. NRC-Webb meeting, Sept. 30, 1975.
87. J. B. Nims, "Fast Reactor Meltdown Experiments," *Proceedings on Breeding, Economics, and Safety in Large Fast Power Reactors* (Argonne National Laboratory, Oct. 1963; ANL-6792, Dec. 1963), pp. 203–31.
88. WASH-1110, *LMFBR Program Plan*, 10 (Safety): 213–35.
89. T. B. Cochran, *The LMFBR: An Environmental and Economic Critique* (Johns Hopkins University Press, 1974), pp. 47–49.
90. PFES, 4: 11.2-11. See also Webb statement in *Hearings*, p. 184.
91. Boudreau and Erdmann, "Autocatalysis," p. 221, table 4.
92. PFES, 2: 4.2-227. The Clinch River Breeder Reactor PSAR proposes to dispense with specific explosion containment measures that would provide some protection against explosion.
93. WASH-1509, pp. 54, 119 (emphasis added). See also *Hearings*, p. 180.
94. For a cogent discussion of this matter, see J. B. van Erp and H. K. Fauske, "Role and Development of LMFBR Protection Systems," *Proceedings of the Fast Reactor Safety Meeting*, pp. 1430–41.
95. PFES, 6: 38–285. See also J. Wright et al., ANL-7120, p. 219.
96. See n. 94.
97. PFES, 2: 4.2-121, 4.2-225.
98. FFTF *Safety Evaluation*, p. 65.
99. Van Erp and Fauske, "LMFBR Protection Systems," p. 1434.
100. WASH-1509, p. 59.
101. L. J. Koch, "EBR-II: An Experimental LMFBR Power Plant," *Reactor Technology* 14, no. 3 (1971): 286–311; ANL/RDP-13 (Jan. 1973), p. 1.1.
102. ANL/RDP-3, p. 1.1.
103. Koch, "EBR-II," pp. 293–95; ANL-5719, *Hazards Summary Report Experimental Breeder Reactor II* (May 1957), p. 106; and Thompson and Beckerley, *Technology*, 1: 160, 163.
104. FFTF *Safety Evaluation*, pp. 39–41; Clinch River Breeder Reactor PSAR, pp. 4.2-1, 4.2-146–50, 4.2-329, 4.3-16, and 4.3-17.
105. Koch, "EBR-II," p. 295 and fig. 11; and PFES (WASH-1535), 2: 4.2-15.

106. R. Carle et al., "Phenix: Status of the Design before Construction," in ANL-7520, *Proceedings of the International Conference on Sodium Technology and Large Fast Reactor Design* (Nov. 7–9, 1968), pt. 2, p. 247.
107. PFES, 4: 11.2-14.
108. Ibid., 3: 6A.1-125–47.

Chapter Eleven

1. As in the Rasmussen Report (final), WASH-1400 (NUREG-75/014) (Oct. 1975), vol. 11, sec. 16.
2. E. A. Wimum and J. M Harrer, ANL-5781, addendum (rev. 1, Oct. 1960), p. 82.
3. Nuclear safety conference, Mass. legislature (Feb. 5, 1976).
4. U.S. Congress, Public Law 93-438, 93d Cong., sess. 207. See also USNRC NUREG-75/018, "Nuclear Energy Center Site Survey" (Mar. 13, 1975).
5. D. F. Cope, "Nuclear Energy Centers: A Prime Element in Reactor Siting," *Nuclear Safety* 16, no. 3 (May/June 1975): 282–90.
6. Based on WASH-740, app. H.
7. PFES, 4: 7.2-15–21.
8. H. C. McIntyre, "Natural-Uranium Heavy-Water Reactors," *Scientific American* 233, no. 4 (Oct. 1975): 17–27.
9. "Pickering Generation Station," brochure, Public Relation Reviews, Ontario Hydro, Toronto, 5th page.
10. Telephone conversation between W. G. Morison, chief design engineer for Pickering Generation Station, Ontario Hydro, and R. E. Webb, Dec. 1975.
11. G. L. Wessman and T. R. Moffette, "Safety Design Basis of the HTGR," *Nuclear Safety* 14, no. 6 (1973): 618. See also WASH-1535 (PFES), vol. 3, sec. 6A.1.
12. *Nuclear Safety* 16, no. 5 (1975): 646, 649.
13. Based on figures given in U.S. Congress, *Hearings on Naval Nuclear Propulsion Program,* Joint Committee on Atomic Energy, 92d Cong., 2d sess. (1972–73), p. 161.

Chapter Twelve

1. "Congressman Mike McMormack Speaks Out on the Impact of Energy on U.S. Jobs," *Nuclear Power Newsletter,* General Electric Co., Nuclear Energy Division, San Jose, Calif. (summer 1975).
2. "Wallace Behnke of Commonwealth Edison Compares Nuclear with Fossil Generation," *Nuclear Power Newsletter* (summer 1974).

Chapter Thirteen

1. Atomic Energy Act, secs. 1, 3.a, and 3.d.
2. Said the present chairman of the U.S Nuclear Regulatory Commission, "I would not have come to this job if I thought that nuclear

power was inherently unsafe." Address to the Edison Electric Institute, 43rd Annual Convention, June 4, 1975, NRC Release no. s-8-75, by W. A. Anders.

3. See sec. 170 of the Atomic Energy Act of 1954, as amended. "Indemnification and Limitation of Liability," and sec. 2.1.
4. U.S. Constitution, Article V.
5. U.S. Congress, Senate, Joint Committee on Atomic Energy, *Amending the Atomic Energy Act of 1946, as Amended, and for Other Purposes*, 83d Cong., 2d sess., June 30, 1954, S. Rept. 1699, p. 10.
6. *Elliot's Debates*, 3: 501. Blackstone's *Commentaries on the Laws of England* (ed T. M. Cooley, 1884), p. v.
7. Blackstone, *Commentaries*, intro., p. 59.
8. See Max Farrand, *The Records of the Federal Convention of 1787*, rev. ed. (Yale University Press, 1937).
9. See e.g. *Elliot's Debates* generally.
10. Max Farrand, *Framing of the Constitution* (Yale University Press, 1913), p. 45.
11. Letters of James Madison to Andrew Stevenson, Speaker of the House, Nov. 27, 1830, in *Letters and other Writings of James Madison*, 4: 120–39.
12. Farrand, *Framing of the Constitution*, p. 216. *The Federalist*, no. 41.
13. *Elliot's Debates*, 3: 466.
14. Ibid., p. 599.
15. See e.g. the letters of Madison to Stevenson, Nov. 27, 1830, in *Letters and other Writings of Madison*, 4: 120–39. See also A. A. Lipscomb, *The Writings of Thomas Jefferson*, 3: 145–53; and *Elliot's Debates*.
16. Farrand, *Records*, 2: 321, 322 (see entry dated Aug. 18, 1787).
17. *The Federalist*, nos. 14 and 45.
18. 206 U.S. 89, 90.
19. U.S. Congress, Senate document no. 154, 68th Cong., 1st sess., p. 81. See also *U.S.* v. *Boyer*, 85 Fed. 432.
20. 297 U.S. 1-78, 78.
21. 297 U.S. 64-68.
22. 297 U.S. 73.
23. *Steward Machine Co.* v. *Davis*, 301 U.S. 586, 587; *Helvering* v. *Davis*, 301 U.S. 619, 640; *Cleveland* v. *U.S.*, 323 U.S. 329.
24. *Helvering* v. *Davis*, 301 U.S. 640.
25. 339 U.S. 725, 738.
26. U.S. Constitution, Article III.
27. James Monroe's "Views on the Subject of Internal Improvements," in S. M. Hamilton, *The Writings of James Monroe* (1822), 6: 216–84. J. Story, *Commentaries on the Constitution of the United States* (1873 ed.), 1: 695–704; see also 1833 ed. 1: 444–56.
28. Monroe's "Message on Internal Improvements," U.S. Congress, *Annals*, May 10, 1822, p. 1803.
29. "Views on Internal Improvements," *Writings of Monroe*, pp. 265, 269.

30. Ibid., pp. 249–50.
31. Alexander Hamilton, "Report on Manufactures," Dec. 5, 1791, *American State Papers*, vol. on Finance, pp. 123, 136.
32. See generally *Elliot's Debates*, New York convention.
33. Letters of Madison to Stevenson, Nov. 27, 1830, *Letters and other Writings of Madison*.
34. Farrand, *Records*, 1: 281–311, for June 18, 1787.
35. Ibid., 3: 627, app. F (the Hamilton Plan).
36. Ibid., 1: 291, for June 18, 1787.
37. Lindley Murray, *English Grammar* (1809), pp. 183–86.
38. See also *The Federalist*, no. 22; Farrand, *Records*, 1: 135, 2: 342–43, which distinguishes "manufactures" and "commerce"; and the various resolutions of the Continental Congress on the want of a power to regulate commerce, which led to the convening of the Federal Constitutional Convention of 1787 (*Elliot's Debates*).
39. 297 U.S. 63–64, 68.
40. 128 U.S. 20–21.
41. 247 U.S. 251, 271–273.
42. 312 U.S. 100, 118 (emphasis added).
43. Chap. 1, sec. 2.c (emphasis added).
44. 9 Wheat. 189.
45. 4 Wheat. 421 (emphasis added).
46. *Elliot's Debates*, 3: 205–7.
47. 317 U.S. 118.
48. *Webb* v. *AEC*, U.S. District Court, Southern District of Ohio, Eastern Division, Civil Action no. 72-14, AEC's motion to dismiss, pp. 6–8. This constitutional issue was submitted to the court, but the court refused to hear the case. *Hearings*, p. 188.
49. 9 Wheat. 189.
50. Blackstone, *Commentaries*, 3: 396.
51. Ibid., intro., p. 71.
52. A definition of the "judicial power" is given in ibid., 3: 24.
53. U.S. Constitution, Article VI.
54. No. 78.
55. See e.g., R. D. Heffner, *A Documentary History of the United States* (Mentor Books, 1965), chaps. 16–23: "The Gilded Age" through "The Roosevelt Revolution."

Appendix One

1. USNRC WASH-1400 (NUREG-75/014), *Reactor Safety Study* (Oct. 1975; final Ras. Rpt.).
2. Ibid., app. 11, sec. 9.
3. See p. 24 herein; also, European Nuclear Energy Agency, Organisation for Economic Co-Operation and Development, *Water Cooled Reactor Safety* (May 1970), pp. 165, 171.
4. ORNL-NSIC-103, p. 11; ROE 69-5.

5. N. J. Palladino, "Mechanical Design of Components," in Thompson and Beckerley, *Technology,* 2: 124–25.
6. Ras. Rpt., draft, app. 6, pp. 9 and 12, BWR 3 release category.
7. Final Ras. Rpt., p. 60, BWR category 2; and app. 6, pp. 2-4 and 2-5.
8. Ibid., app. 11, p. 6-2.
9. Ibid., app. 6, p. 8-14, 8-16, E-31 through 38. Compare with WASH-740, pp. 41 and 42.
10. D. R. Inglis, *Nuclear Energy: Its Physics and Its Social Challenge* (Addison-Wesley, 1973), pp. 128–30.
11. WASH-740's value is based on reducing .215 μCi/m^2 by a factor of 2.8 to provide a safety factor, though "very little" (p. 41).
12. FRC Report no. 7, May 1965, cited in *Radiological Health Data and Reports* 10, no. 7 (July 1969): 301.
13. U.S. Environmental Protection Agency, *Environmental Radiation Protection Requirements for Normal Operations of Activities in the Uranium Fuel Cycle,* draft environmental statement for a proposed rulemaking action (May 1975), p. 65.
14. Inglis, *Nuclear Energy,* p. 131.

Appendix Two

1. R. O. Brittain, "Analysis of the EBR-I Core Meltdown," *Proceedings of the International Conference on the Peaceful Uses of Atomic Energy* (Geneva, 1958). Thompson and Beckerley, *Technology,* 1: 628. G. S. Lellouche, "The EBR-I Incident: A Re-examination," *Nuclear Science and Engineering* 56, no. 3 (Mar. 1975): 303–7. Webb, "Some Autocatalytic Effects," pp. 22–23.
2. Scott, "Fuel Melting Incident," pp. 123–34. D. B. Wehmeyer and C. S. Yeh, "Reactivity Changes in Meltdown Configurations of Fermi Core A Fuel," APDA-LA-1 (June 1969). Webb, "Some Autocatalytic Effects," pp. 127–37, 175–96. H. J. Gomberg et al., "Report on the Possible Effects on the Surrounding Population of an Assumed Release of Fission Products into the Atmosphere from a 300-Megawatt Nuclear Reactor Located at Lagoona Beach, Michigan, APDA-120 (July 1957).
3. Personal observation as an employee of the Division of Naval Reactors, USAEC; see also AEC Authorization Legislation FY1966, U.S. Congress, *Hearings before the Joint Committee on Atomic Energy,* 89th Cong., 1st sess. (Feb. and Apr. 1965), pt. 2, pp. 1268–72; *Shippingport Pressurized Water Reactor* (Addison-Wesley, 1958).
4. G. R. Gallagher, "Failure of N Reactor Primary Scram System," *Nuclear Safety* 12, no. 6 (Nov./Dec. 1971): 608.
5. USAEC WASH-1270, "Anticipated Transients without Scram for Water-Cooled Power Reactors" (Sept. 1973), p. 49. Letter from S. H. Hanauer, Director, Office of Technical Advisor, Department

of Regulation, USAEC, to R. E. Webb, dated Dec. 9, 1974. Reactor Operating Experience report, USAEC ROE-71-16.

6. Consumers Power Co., *Consumers Power News*, Nov. 1967 (Jackson, Michigan), p. 20; *Detroit News*, Jan. 9, 1971, p. 3.
7. *Nuclear Safety* 15, no. 2 (Mar./Apr. 1974): 210–11. Letter from Vermont Yankee Nuclear Power Corp. to AEC, Nov. 14, 1973 (VYV-3071, AEC docket no. 50-271, pp. 1–6).
8. See ORNL-NSIC-103, "Abnormal Reactor Operating Experiences, 1969–1971," ROE-71-2.
9. Ibid., "Unscheduled BWR Blowdown," ROE-71-4.
10. *Nuclear Safety* 16, no. 4 (1975): 510. D. D. Comey, "The Incident at Browns Ferry," in Friends of the Earth, *Not Man Apart*, Sept. 1975. "Browns Ferry Nuclear Plant Fire," U.S. Congress, *Hearings before the Joint Committee on Atomic Energy*, 94th Cong., 1st sess. (Sept. 16, 1975), pt. 1, pp. 73–78, and app. 7 and 15. Ras. Rpt., WASH-1400 (draft), 3: 71. ORNL-NSIC-103, pp. 7-10 and 9.
11. *Nuclear Safety* 16, no. 3 (May/June 1975): 378. "Nuclear Regulatory Commission Action Requiring Safety Inspections Which Resulted in Shutdown of Certain Nuclear Power Plants," U.S. Congress, *Joint Hearing before the Joint Committee on Atomic Energy and the Committee on Government Operations of the U.S. Congress*, 94th Cong., 1st sess. (Feb. 5, 1975).
12. J. M. Miller, "Incident at the Lucens Reactor," *Nuclear Safety* 16, no. 1 (Jan./Feb. 1975): 76–79. *Nuclear Safety* 16, no. 4 (July/Aug. 1975): 511. USAEC ROE-70-4, "Loss of Coolant Accident," ORNL-NSIC-103, pp. 66–68.
13. Thompson and Beckerley, *Technology*, 1: 653.
14. Ibid., pp. 633–36. H. J. Dunster et al., "District Surveys Following the Windscale Incident, Oct. 1957," p/316 UK, sess. D-19, *International Conference on the Peaceful Uses of Atomic Energy* (Geneva, 1958).